COMPARATIVE BREEDING BEHAVIOR OF FOUR SPECIES OF NORTH AMERICAN HERONS

PUBLICATIONS OF THE NUTTALL ORNITHOLOGICAL CLUB

NO. 2

COMPARATIVE BREEDING BEHAVIOR OF FOUR SPECIES OF NORTH AMERICAN HERONS

by

ANDREW J. MEYERRIECKS

CAMBRIDGE, MASSACHUSETTS

Published by the Club

1960

Publications of the Nuttall Ornithological Club

1. *The Comparative Biology of the Meadowlarks (Sturnella) in Wisconsin.* Wesley E. Lanyon 1957. [viii] + 67 pp., 31 pl.
2. *Comparative Breeding Behavior of Four Species of North American Herons.* Andrew J. Meyerriecks 1960. viii + 158 pp., 15 pl.

Printed in the United States of America

Allen Press • Lawrence, Kansas

Publications of the Nuttall Ornithological Club may be obtained from the Massachusetts Audubon Society, Drumlin Farm, South Lincoln, Massachusetts.

ACKNOWLEDGMENTS

I am grateful to Professor Ernst Mayr, my graduate sponsor, for his assistance in every phase of this study. For help in the field, I should like to thank Robert Porter Allen, Robert Meyerriecks, Herbert Johnson, Charles M. Brookfield, and Miguel Alvarez del Toro. For the loan of photographs, I am grateful to Andreas Feininger (Plates 1, 3, 4, and 6), Robert Meyerriecks (Plates 2, 5, and 7), and Samuel A. Grimes (Plate 15). The remaining photographs were taken by the author. Charts and maps were prepared by Robert and George Meyerriecks.

Field work was supported by grants from the Society of Fellows, Harvard University, and by Grants-in-Aid, Sigma Xi-RESA.

Mrs. Nancy Buffler, of the Museum of Comparative Zoology, Harvard College, deserves special thanks for her excellent line drawings of herons.

I wish to thank the following persons for their suggestions and critical reading of the manuscript: Professor Ernst Mayr, Dr. Oliver L. Austin, Jr., Dr. William H. Drury, Jr., Dr. P. Marler, and Oscar M. Root.

Finally, I am particularly indebted to my wife, Norma, for help in many aspects of this work.

ANDREW J. MEYERRIECKS
Assistant Director
Hatheway School of
Conservation Education
South Lincoln, Massachusetts

CONTENTS

INTRODUCTION

The herons, egrets, and bitterns of the family Ardeidae are, for various reasons, particularly suitable subjects for a comparative ethological study. The family includes approximately sixty-three species, grouped in fifteen genera by the latest reviser (Bock, 1956). Study of geographical variation of individual behavior traits, a neglected aspect of ethology, is facilitated by the broad distributions of many species in the temperate and tropical regions of both hemispheres (for example, the Black-crowned Night Heron, *Nycticorax nycticorax* (Linnaeus)). The local abundance of many species affords an opportunity for the repetition so necessary in observations of behavior patterns.

Herons and their allies possess a variety of modified plume types, and the existence of such structures readily leads to a study of their use as isolating mechanisms during pair formation in mixed colonies. Herons perform highly ritualized ceremonies during pair formation and nest relief, thus affording an opportunity for comparisons of species and genera. Within the family, breeding behavior ranges from solitary nesting (American Bittern, *Botaurus lentiginosus* (Rackett)) through loose aggregations (Green Heron, *Butorides virescens* (Linnaeus)) to highly social nesting (Snowy Egret, *Leucophoyx thula* (Molina)), thus facilitating the study of the origin and evolution of social nesting habits.

Polymorphism, in the form of color phases, is common among ardeid birds (e.g., Reddish Egret, *Dichromanassa rufescens* (Gmelin), and Reef Heron, *Demigretta sacra* (Gmelin)), and the occurrence of such phases readily leads to a study of their selective and potential values as isolating mechanisms during pair formation.

Bock based his generic revision of the family primarily on morphological characters; a comparative behavior study might lead to a more satisfactory classification using both behavioral and structural characters. In addition, Heinroth (1929), O. von Frisch (1957), and others have shown how easily herons may be raised in captivity, a welcome adjunct to field observations.

Unfortunately, the student of heron behavior lacks the rich literature available to the student of gull or stickleback behavior. With the exception of the valuable papers of Verwey (1930), Lorenz (1938), Noble *et al.*, (1938, 1940, 1942), and a few others, the available literature on heron behavior is scattered and cursory. Hence, the conclusions presented here are to be considered first approximations.

Prior to any comparative study, however, an attempt must be made to study intensively one species in order to obtain a comparative standard for the group. The Green Heron was selected for this purpose, not only because it is a widespread and locally abundant species, but also because little was known about the general reproductive biology of the bird. Such information would naturally accumulate during the course of the general ethological investigation. Accordingly, I began a study of the Green Heron in the spring of 1954 and continued research during the breeding seasons of 1955 and 1957. However, this research did not result in the accumulation of large amounts of quantitative data. Although Green Herons are locally common and may breed in small to large colonies, the actual number of breeding pairs is not great when compared with other more highly social species of herons such as the Snowy Egret. In addition, the reed-choked nesting areas prevented observation of more than a few birds at any one time; hence, some of the conclusions presented here are based on relatively few repeat observations. Study of a more highly social species such as the Snowy Egret, which breeds in large numbers along the Gulf Coast of North America, might lead to quantitative results rivaling those gathered in certain European gulleries.

Part I of this paper presents the Green Heron study. Part II deals with the Great White-Great Blue Heron, Reddish Egret, and Snowy Egret. Some evolutionary trends in heron behavior are discussed in Part III. A comparative behavior chart follows Part III; it is designed to show at a glance the existing state of knowledge of the behavior of most North American herons.

This paper is part of a Doctoral Dissertation presented to the Faculty of the Biological Laboratories, Harvard University, in partial fulfillment of the requirements for the degree of Doctor of Philosophy, May, 1958.

I have used, with few exceptions, the scientific and common names listed in *Check-List of North American Birds,* American Ornithologists' Union, 1957.

Part I

THE BREEDING BEHAVIOR OF THE GREEN HERON

***Butorides virescens* (Linnaeus)**

ORGANIZATION

Part I presents a preliminary description and analysis of the reproductive behavior of the Green Heron. I have arbitrarily limited the description to that portion of the breeding cycle which extends from the time the birds arrive on the breeding grounds until the first egg is laid. I have tried to describe the behavior of the Green Heron in a chronological order, but so much of the reproductive behavior, especially the early phases, involves hostile (attack and/or escape) behavior, that a special section is devoted to describing it before the description of pair formation and related activities. For comparative purposes, a brief introductory section describes flight, preening, and other maintenance activities.

METHODS

I made field observations of the Green Heron during one partial season (1954) at breeding sites in eastern Massachusetts (Wayland, Burlington, and Newburyport) and western Long Island, New York, and during two complete seasons (1955, 1957) in colonies on Rulers Bar Hassock, an island in Jamaica Bay, Long Island, New York (Figure 1). I made additional observations in Florida Bay in the winter of 1955–56 and a few near Puerto Arista, Chiapas, Mexico, in the summer of 1956. Most of the observations I made from blinds with binoculars (7×) and a telescope (20×). The drawings illustrating the text were made from my own field sketches and from photographs. A conservative estimate of the time I spent in the field observing the Green Heron is 3,000 hours.

STUDY AREA

Most of the field work was concentrated on Rulers Bar Hassock, a salt marsh island about three miles long and one-half mile wide in the middle of Jamaica Bay. The town of Broad Channel occupies the lower third of the island, and a main thoroughfare, Cross Bay Boulevard, runs down the central axis. I found four separate heron breeding sites, two east of the boulevard and two west ot it (areas 1, 2, 3, and 4, Figure 1). All four sites contained reeds, *Phragmites communis*; bayberry, *Myrica carolinensis*; poison ivy, *Rhus radicans*; wing-rib sumach, *R. copallina*; staghorn sumach, *R. typhina*; cottonwood poplar,

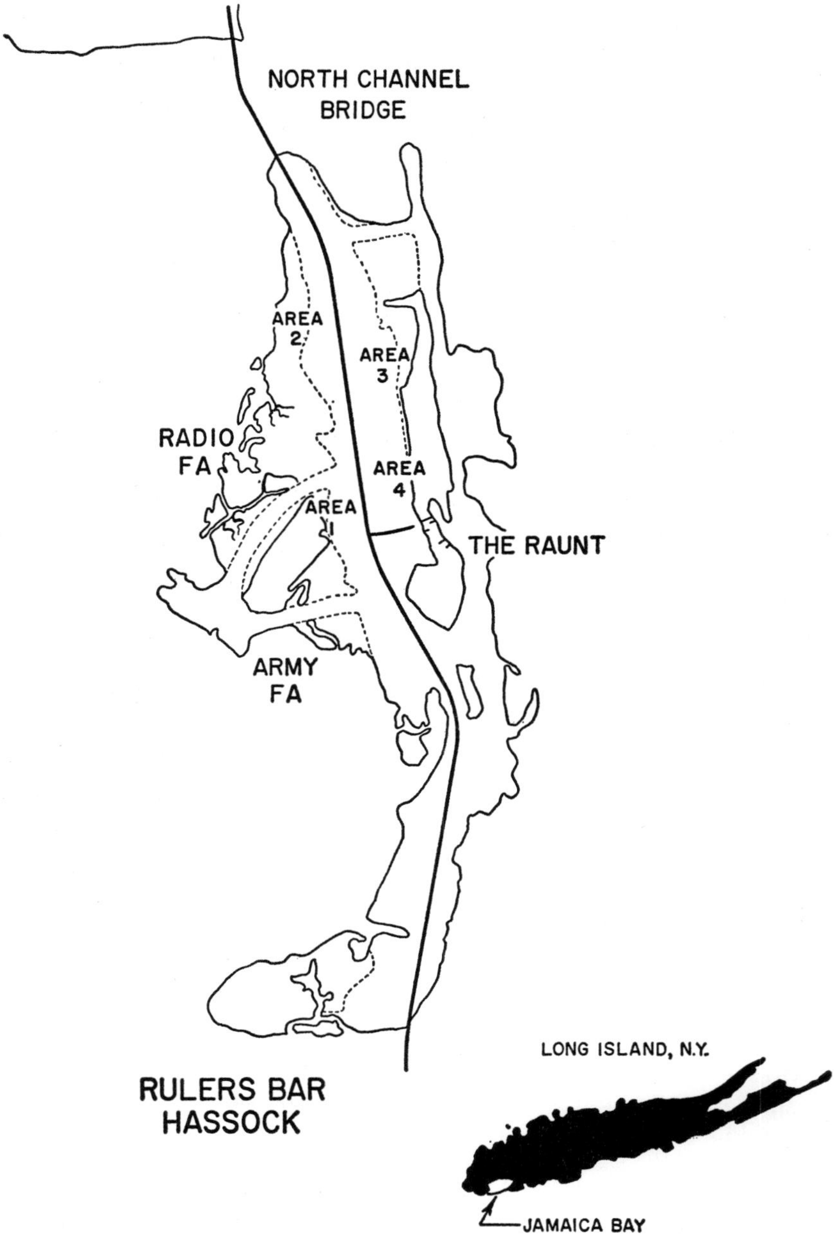

FIGURE 1. Map of Rulers Bar Hassock, an island in Jamaica Bay, western Long Island, New York. "FA" stands for feeding area.

Populus deltoides; and willow, *Salix discolor*. The herons fed in the extensive marshes surrounding the breeding areas. In 1955 the largest number of Green Herons bred in area 1 (70 pairs), somewhat fewer in area 2 (41 pairs), and scattered pairs in areas 3 (17 pairs) and 4 (8 pairs). I made observations in all four areas throughout the 1955 and 1957 breeding seasons, but I recorded most of the data in area 1.

DISTRIBUTION AND DESCRIPTION

The Green Heron, a highly migratory species, is widely distributed as a breeding bird in North and Central America, ranging from western Washington, southern Ontario, and southern New Brunswick in the north, south through Mexico, Central America, and the West Indies to the Pearl Islands in the Gulf of Panama. It breeds in a few localities in northern South America, where it meets the closely related Striated Heron, *Butorides striatus* (Linnaeus), of South America, Africa, and Austral-Asia. Another close relative is the Galapagos Heron, *Butorides sundevalli* Reichenow, restricted to the Galapagos Archipelago.

These small herons may nest solitarily, in loose aggregations, or rarely in colonies. They are locally common breeders in the United States near fresh and salt water.

The Green Heron, about eighteen inches long, is among the smaller members of the family Ardeidae, with an adult weight of approximately 200 grams (Slipp, 1942; Meyerriecks, unpub. MS). Haverschmidt (1948b) records the weights of two adult Striated Herons from Surinam as 173 and 187 grams.

In the field, Green Herons appear as small, dark, short-legged herons with rather shaggy crests. The back plumage often appears to be more blue than green. The plumage of the two sexes is similar, with dark, greenish-black crests, dark green nape and back, chestnut-maroon neck and brownish underparts striped with white. The green, lanceolate plumes of the scapular region reach the tail. The nuptial coloration of the male is full, intense, and lustrous, whereas that of the female is noticeably less brilliant. Hindwood (1933) states, for *striatus* during the breeding season, "although the female is said to be similar to the male, it always seemed to me when I had both birds in view at the same time that she was duller in the colouration of her plumage." I agree emphatically with respect to *virescens*.

MAINTENANCE ACTIVITIES

Certain movements and behavior patterns that occur throughout the life of the bird must be known to understand the specialized displays, especially of the reproductive period, that are derived from them.

Locomotion

Flight. The flight is similar to that of the Common Crow, *Corvus brachyrhynchos* Brehm, but it is more direct and generally unhurried. The wing beats appear to be more rapid, the wings are arched in flight, and the bird's body seems to rise and fall with every flap. Blake (1948) records 2.8 flaps per second in normal flight, the rate increasing to 3.8 when the bird hurried. McLean (1930) gives the speed of a Green Heron in "steady, easy flight" as 34 mph. The bird was flying at an altitude of twenty-five feet, and McLean timed the bird with an automobile. Wood (1933) gives a flight speed of 22 mph. I timed a Green Heron flying north on Cross Bay Boulevard on April 20, 1955. The wind was about 10 mph (Beaufort 3) from the northeast. For three-fourths of a mile the bird kept pace with my automobile at a ground speed of 25 mph.

On long flights the neck is retracted, and the legs and feet are held to the rear in line with the body. However, on short flights the bird keeps its head and neck extended, the head slightly above the level of the back. During the breeding season, this latter type of flight is very common on short and long flights. In Pursuit Flight (p. 39) the bird being chased flies with its head and neck held well above the level of its back, and watches its pursuer intently.

Two variants of normal flight are conspicuous during the breeding season: (1) the Flap Flight, a sexual display with hostile overtones (see p. 41), and (2) a direct, rapid flight for attacking another bird either perched nearby or passing in flight.

Green Herons occasionally glide just before landing; they almost never glide during a sustained flight. Only once have I seen this species perform the spectacular spiral-descent type of flight so often displayed by other herons. A Green Heron flying over the Waters River near Danvers, Massachusetts, on June 6, 1954, suddenly started to drop toward the river in a whirling, twisting fashion, with its feathers, especially the scapular plumes, streaming in the wind. The bird abruptly checked its spiral descent when it was a few feet from the shore of the river, where it landed and proceeded to feed.

Taking off. The actions of a bird about to take flight are of significance as intention movements. As long ago as 1910 Oskar Heinroth showed that certain bird movements give the observer a clue to what the bird intends to do. Heinroth called these incomplete or preparatory movements, especially those involved in locomotion, *Intentionsbewegungen* (intention movements). (See Daanje (1950) for a thorough review of the function and evolution ("ritualization") of intention movements in birds.) Many displays appear to be derived from the basic movements of walking and intention movements such as tail-flipping.

An accurate and complete description of all the movements performed by a Green Heron while taking off requires high-speed photography. The following actions are based only on field observations and ordinary still photographs:

1. all feathers are sleeked
2. legs are bent
3. body is brought to a horizontal posture, wings slightly extended
4. head and neck are partially retracted
5. tail appears to be raised and fanned laterally, or it may be flipped
6. body is thrust forward and upward by extension of the legs
7. wings are extended upward
8. tail is depressed
9. head and neck are extended, crest may or may not be erected
10. legs and feet are brought together and extended horizontally rearward
11. head and neck are withdrawn to the body.

Green Herons perched on the ground in a wooded area characteristically walk to a nearby shrub and then make very short jumps to the top of the shrub before taking off. If they are perched in the lower branches of an open tree, they usually climb to the crown branches before flying.

Numerous variations occur in this typical sequence. For example, a bird perched in a vegetation tangle jumps from a semivertical rather than a horizontal stance. Retraction of the head and neck may begin from an alert posture. Still photographs show that some Green Herons extend the legs rearward at once, while others delay this movement until after the head and neck are drawn back to the body.

Landing. As with taking off, an exact description of landing is difficult without the aid of high-speed photographs. However, the Green Heron appears to execute the following movements:

1. short glide, legs and feet trailing, toes flexed, head and neck retracted (glide may be omitted)
2. head and neck extended and elevated, crest erected (almost without exception)[1]
3. legs dangled, then thrust forward and downward
4. tail spread
5. wings beat as if hovering
6. perch is grasped, head and neck retracted, crest lowered, body in upright position; or, head and neck extended, crest lowered, body upright.

Walking. "Sneaking" is a more descriptive term than "walking" for the Green Heron. Before starting, the bird follows steps 2, 3, and 4 of the preparation for flight: it bends the legs slightly, inclines the body forward, and retracts the head and neck. Then the bird slowly brings one leg forward, toes relaxed; gradually it brings the body forward as the foot with toes widespread touches the ground. The bill and head keep in line with the crouching body. The bird's every action is deliberate and reminiscent of a cat hunting.

[1] Glenister (1951) and Smythies (1953), among others, remark that the Striated Heron, *B. striatus*, usually erects its crest just before landing.

FIGURE 2. Green Heron showing Stand and Wait feeding behavior.

FEEDING BEHAVIOR

Feeding behavior is primarily of two types, Stand and Wait and Wade or Walk Slowly. I shall also describe several other interesting but relatively unimportant types.

Stand and Wait. The Green Heron stands motionless, head and neck retracted, body held either low and horizontally (Figure 2) or semivertically. When striking from the horizontal position the bird darts its head and neck forward and down; if striking from the semivertical position it shoots the head and neck out and down, at the same time lowering its body to the horizontal position. Green Herons feed this way most commonly in shallow water on the shore of a lake or stream. A common variant is waiting while perched on a post (Stone, 1937), a limb, stub, or log (Saunders, 1926), or the transom of a rowboat. Warburton (1948) watched a Green Heron perched on a log catching flying dragonflies. When using such a perch, the hunting bird may either crouch with head and neck retracted, or may wait with its head and neck extended and down, peering into the water. The short legs prevent wading in deep water; it fishes in deep water by perching on logs and prop roots of Red Mangrove, *Rhizophora mangle,* and by diving.

Wade or Walk Slowly. This is very similar to walking. As the bird stalks closer to its prey, its steps become slower and longer; each foot is brought forward, placed, and then lifted so slowly that the movements are barely perceptible. The entire tarsus may almost touch the ground. The bird may retract its head and neck or hold them extended over the water or ground; or rarely the head and neck are held momentarily in an extended "peering over" attitude. The strike is very swift. I have never seen a Green Heron seize its prey with any but a scissor-like grip; that is, the prey is not pierced by the bill.

Diving. Catching prey by diving into the water from a plank (Barker, 1901), the stone edge of a park pool (Brooks, 1923), or from the shore of a pond or stream (Kalter, 1932; Hawbecker, 1949) are fairly common variants of normal feeding behavior. Brown (1949) observed *B. striatus* diving for food.

Once caught, the prey is held crosswise between the tips of the mandibles, and then it is tossed head first into the gape. Characteristically, Green Herons dip their bills in the water and shake their heads after swallowing aquatic prey. The bird lowers the forward part of its body toward the water and dips its bill vertically into the surface, raises its head and bill clear of the water and then shakes the bill vigorously. Kalter (1932) has also observed bill-dipping by the Green Heron after swallowing prey. On many occasions, I have seen Green Herons dip their bills after they had missed a strike.

Care of the Body Surface

Preening. Green Herons can reach their entire body surface with their bill except the top of the head and the upper part of the neck. The sequence in which the feather tracts are preened varies with the individual, but there is almost no variation in the movements used. Some individuals preen first the entire right side and then the left, others rapidly alternate right-left-right. Most birds begin by preening the breast feathers and end by oiling the bill on the oil gland.

The following preening sequence is fairly typical: the bird first nibbled the breast powder down patch and then the breast feathers. It grasped each feather at the base in the tips of the mandibles and pulled up and out with short, smooth movements. Next the heron nibbled the breast powder down patch, pulled additional breast feathers out, nibbled several, and then partly opened one wing and put its head and bill between the wing and body. Now the bird shook and rubbed its head vigorously against the body. It then preened its abdominal feathers with rapid, flicking movements of the head, each feather being grasped momentarily, then released with a short, outward flick. Abdominal feathers were nibbled for several minutes. Flank feathers were barely touched with the bill, but occasionally one was flicked out. The breast powder down patch was nibbled again, and by this time the bill and fore part of the head were covered with down, which gave these parts, especially the bill, a bluish-white cast.

The bird next lifted the long, lanceolate scapular plumes singly, and either pulled each one entirely through the tip of the bill, or merely flicked one feather upward and released it with a rapid toss of the head. To preen its neck feathers, the bird assumed a rather grotesque attitude (Plate 1). After erecting all the neck feathers, it nibbled individual ones as it drew them out rather than drawing them through the tips of the bill in one smooth stroke.

The bird nibbled its primaries over their entire length or drew them out in one smooth stroke, and it usually preened the primaries while the wing was closed or only partly opened. It flicked primary coverts upward and then released them. Underwing preening was accomplished by fanning one wing, then inverting the head and lightly touching the underside of certain primaries and secondaries. Plate 2 shows a Green Heron preening its underwing. The exact function of this is obscure. The delicate manner in which the tip of the bill barely touches certain feathers suggests the picking off of parasites, but I have never seen anything swallowed at this time.

The last act of preening usually is the rubbing of the bill over the oil gland. The bill is black during the courtship phases of the breeding cycle, and the oiling makes it glisten.

One complete preening bout recorded on 16mm film is as follows: bird nibbled breast powder down patch—pulled out breast feathers—nibbled breast feathers—nibbled breast powder down—placed head under left wing—pulled out and nibbled abdominal feathers—nibbled breast powder down—pulled out breast feathers—placed head under right wing—nibbled breast—flicked scapulars—pulled out abdominals—flicked scapulars—nibbled upper neck feathers—nibbled scapulars—flicked abdominals—upper neck—scapulars—lower neck—scapulars—lower neck—scapulars—lower neck—abdominals—upper neck—flanks—abdominals—head under left wing—upper neck—scapulars—primary coverts—primaries—under wings—abdominals—nibbled flanks—abdominals—bird flew.

Hindwood (1933) gives a good description of powder down patch nibbling and preening in Australian *B. striatus.*

Scratching. The Green Heron scratches by bringing the foot directly up to the head; the wing is not drooped ("vornherum" of Heinroth, 1930; "directly" of Simmons, 1957). Scratching, normally prolonged and vigorous, appears to be done mainly by the extended middle toe, that is, the one bearing the pectinated claw (field observations supported by 16mm motion pictures). During all my observations of scratching, the bird's crest was erected (see Plate 3). Hindwood (*op. cit.*) noted that *B. striatus* in Australia characteristically erects its crest when head-scratching. With the crest fully erect, a Green Heron slowly rotates its head while the middle toe works the entire surface. Head-scratching may occur at any time during a preening bout; normally the bird scratches its head at least once during a prolonged bout.

Shaking. Occasionally during a preening bout, especially near its end, the bird leans forward, extends its wings about half way, and shakes out all its plumage (Plate 4). It erects all feathers and moves the wings in and out vigorously. When the bird has finished, it sleeks all of its feathers and then rapidly preens a few primaries with swift, smooth strokes. Shaking out is at times a prelude to flight; shaking out in flight after take-off is rare in the Green Heron.

Bill-cleaning. After preening, the bird's bill may have one or more tiny feathers stuck to the tip. The bird removes these with the pectinated claw of the middle toe by lifting its foot upward and forward, extending the middle toe, and flicking the fluff from the bill. The powder down that covers the bill occasionally during prolonged preening is usually brushed off before the end of the bout; if it is not, the oiling of the bill removes any remainder.

Bill-wiping. Green Herons wipe their bills by stropping them on a branch. The bird lowers its head, bill vertical, and rubs the bill on the branch, alternating left-right-left with swift and smooth strokes.

Pecking the feet. At any time during a preening bout the bird may reach down and peck its toes with the tip of the bill. As far as I can tell, the bird does not touch the tarsi.

Stretching. In their usual stretch motion, Green Herons bend their body forward, extend their head and neck, shift their weight to one leg, and then stretch the wing and leg of the opposite side. They then shift their weight to the other side and stretch the other wing and leg. They may stretch their wings and legs at any time, but usually during a preening bout.

In another type of stretching, seen more often in nestlings than in adults, the bird stretches its head and neck fully forward, then raises both wings over the body so that they meet.

Gagging and bubbling. After feeding, Green Herons often raise their head, open their bill and make one or more swallowing movements. Unlike the normal movements of swallowing, the bird appears to be gagging as if something were caught in its throat, and the violent movements apparently remove the obstruction. Adults feeding a brood characteristically bubble their gular region by rapidly moving the gular area in and out. This seems to facilitate disgorgement of the food. Adults may follow gagging with bubbling movements.

Sun-bathing. Hauser (1957) lists the Green Heron as a "voluntary" sun-bather, but she does not illustrate or describe the position. When sun-bathing, these herons open their wings about halfway and let them droop, raise their heads and necks slightly, and usually face the sun (Figure 3). They may open their bills and elevate various feather tracts, but they show much individual variation. Sun-bathing birds may close their eyes, and they are then easily approached.

Defecation. The defecating bird lowers the back end of its body,

FIGURE 3. Green Heron sun-bathing.

elevates its tail, slightly opens and droops its wings, and then raises its head and neck slightly. The female makes the same movements during egg-laying. Green Herons frequently defecate when they are frightened, usually right after take-off. This usually occurs in high-intensity escape situations when the bird has little or no warning of the approach of the intruder, and is part of a general fright response. When they have ample warning, Green Herons seldom defecate on take-off.

Sleeping. A sleeping individual usually has the neck fully retracted, wings slightly drooped, body feathers slightly erected, and eyes closed. Birds sleeping on the ground frequently raise one leg so that only the foot is visible protruding from the fluffed feathers. They frequently sleep while incubating and brooding (Plate 5) and the head may rest on the rim of the nest. In this position, the bird usually closes its eyes, but sometimes only the nictitating membrane. The lids are closed for varying periods.

Yawning. This movement occurs most frequently after long periods of incubation, and may indicate an oxygen need. Yawning seems to be one of the general comfort activities (stretching, walking, preening) following long periods of inactivity while sitting on the eggs.

Relaxation. Often Green Herons adopt the normal perch position (Plate 6) and spend long periods without moving. The bird appears to be relaxed, no special movements occurring other than an occasional glance at a passing bird. When perched low in protective cover, the relaxing bird may cover its eyes with the nictitating membranes for several seconds at a time.

Alarm, "Freezing," and Escape Responses

Responses to danger depend on the type and distance of the stimulus source. Responses may be grouped into three categories: (1) flight; (2) alert responses; and (3) Bittern Stance. The Bittern Stance is never accompanied by a call, but the others may be associated with one or more calls.

Flight. If a Green Heron is fully aware of the approach of a human, it usually flies away unhurriedly. Such a flight is accompanied by a single, rather high-pitched "skyow" or "skeow," the typical call of the species. The flight of a startled bird is usually hurried, with one or more "skyow" calls. Precipitous flights are invariably accompanied by one or more high-pitched "skyows." If a Marsh Hawk comes too close, the heron flies off unhurriedly, sometimes with several "skyow" calls.

Incubating or brooding birds usually rise from the nest and walk rapidly to the top of the nest tree when a human approaches slowly. They may then fly to a nearby tree, or rarely out of sight. Flight is unhurried but usually accompanied by "skyows." If the intruder remains quiet or hides, the bird will fly back to the vicinity of the nest, but as soon as it sees the intruder, it explodes into flight, giving a rapid burst of very high-pitched "skyows," low, raspy "skuk-skuk" or "kuk-kuk" calls, or high-pitched, rasping "raaah-raaah" calls. These calls, especially the last, are repeated until the intruder goes away. The "raaah" call expresses stronger hostile tendencies than does the "skuk-skuk" call.

If an intruder surprises the bird on the nest, the bird flies off explosively giving "skuk-skuk" or "raaah-raaah" calls, lands nearby, and keeps calling. Green Herons do not hover over the nest or the intruder as Black-crowned Night Herons do. If the visitor stays quiet, the heron starts to look for him.

Alert postures. When they have ample warning of approach, Green Herons take one of several alert postures (Figure 4), varying in the amount of stretching of the neck, flipping of the tail, and/or raising of the crest.

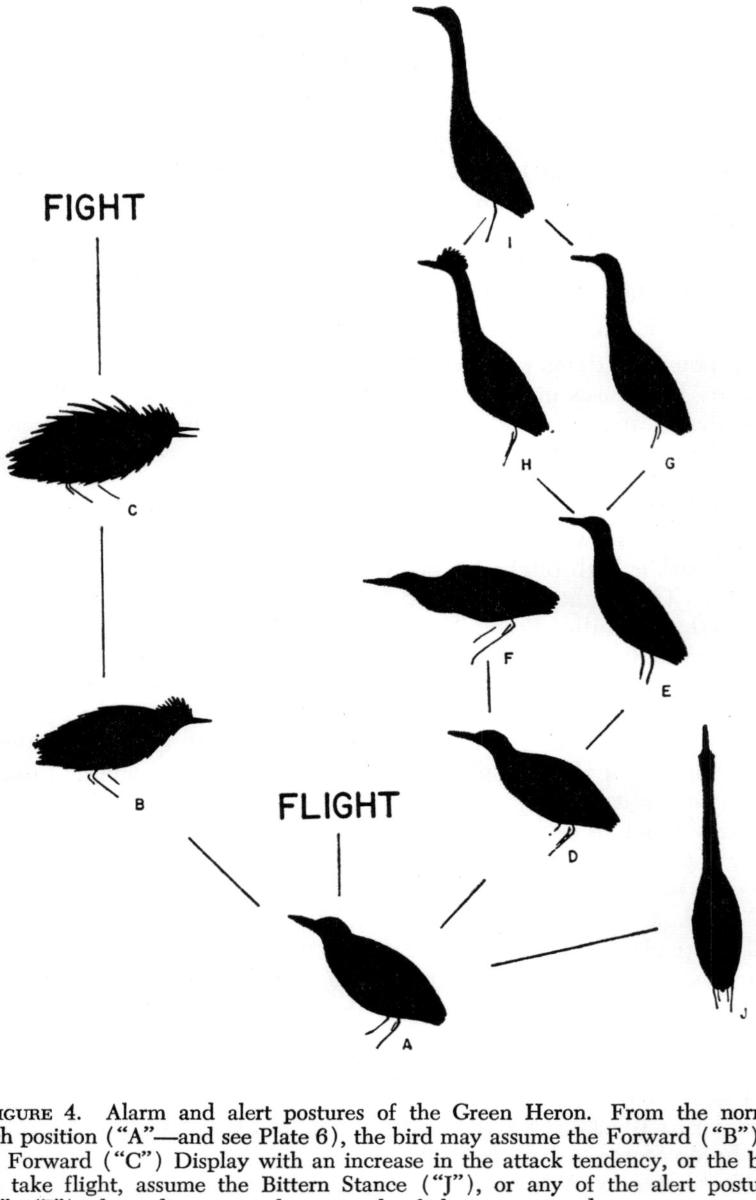

FIGURE 4. Alarm and alert postures of the Green Heron. From the normal perch position ("A"—and see Plate 6), the bird may assume the Forward ("B") or Full Forward ("C") Display with an increase in the attack tendency, or the bird may take flight, assume the Bittern Stance ("J"), or any of the alert postures ("D"—"I"), depending upon the strength of the escape tendency.

Plate 6 shows the normal perch position; alert postures are modifications of this. E in Figure 4 shows a bird in the normal alert posture. I believe that the extension of the neck reflects the strength of the escape tendency—the greater the escape tendency, the greater the extension of the neck. Alert postures show components of preparation for flight (see list, p. 7), specifically feather-sleeking, leg-bending, and tail-flipping. From any alert posture, forward inclination of the body and slight neck retraction prepare the bird for flight. I, Figure 4, is an extreme alert posture adopted by birds attempting to locate a predator hidden in a tangle some distance away. It is an extended alert posture.

When the predator is close to a perched bird, the latter may adopt a horizontal alert posture (F, Figure 4). This posture has the elements of flight preparation and adds rapid extension and withdrawal of the neck as the bird attempts to locate the hidden predator. A bird in these two inquisitive alert postures appears very tense, and it raises its crest and flips its tail conspicuously. Crest-raising in some alert postures indicates a conflict between the tendency to escape and the tendency to remain. I doubt that crest-raising in alert postures is a hostile component because the crest remains fully erect in hostile displays, while in alert postures, the bird rapidly raises and lowers its crest—up-down-up-down. Crest-raising is the most common of all the feather movements of the Green Heron with the exception of wing movements during flight.

Bittern Stance. Green Herons show the Bittern Stance (*Pfahlstellung* of Portielje, 1926) in response to: (1) the sudden appearance of a Marsh Hawk; (2) the alarm calls of several species of birds given suddenly and very close to the Green Heron; and (3) the sudden appearance of a human who surprises an incubating or brooding bird. Three excerpts from my notes demonstrate these three situations:

> May 1, 1954—0937—Green Heron, with brilliant soft-part coloration, perched in sumach ten yards from blind—preening slowly.
>
> 0938—A male Marsh Hawk suddenly swept over the tree, and the heron immediately assumed a full "bittern stance"—when the hawk flew on, the heron relaxed and resumed its preening.
>
> May 5, 1955—1138—Green Heron preening about 15 yards south of blind—male Redwinged Blackbird, perched about 10 yards from the heron and not visible to the latter, suddenly gave one loud "check," and the heron assumed the "bittern stance" at once—the heron relaxed a few moments later and resumed preening.
>
> August 4, 1955—1630—checked nest 1-66 (nest number 66 in breeding area 1)—no adult present—three newly hatched young in nest—I took position in tangle of vines near nest—
>
> 1635—adult returned to nest 1-66—started to brood young at once—I waited for bird to get settled, then suddenly stood up—brooding bird immediately assumed the "bittern stance" still brooding the young!—held this position for one full minute—flushed wildly when I made sudden move (Figure 5).

FIGURE 5. Green Heron in the Bittern Stance while brooding young.

Several heron species show variations of the Bittern Stance, but none, to my knowledge, is so strikingly similar to that of the *Botaurus-Ixobrychus* bitterns as *Butorides*. Cowles (1930) observed a Bittern Stance at the nest of a Striated Heron, *B. striatus*, in Africa.

The Green Heron's Bittern Stance (Figure 6) involves:

1. immobility—except for occasional head-turning (see number 6), the bird remains motionless until the predator leaves or forces the bird to fly
2. body held vertical—the bird holds its head and neck in line with the body; the tail is depressed and the neck is stretched to its maximum and pointed straight upward
3. legs straight—instead of flexing the legs as in flight intention movements (p. 7), the legs are extended
4. feathers sleeked—the bird sleeks its feathers to the extreme
5. eyes bulged—the bird's eyes appear to bulge from its head, and the iris becomes a deep orange
6. head-turning—if the predator is distant, the bird may turn its head very slightly in order to follow the predator's path.

The Bittern Stance components eliminate, reverse, or highly modify a number of the flight preparation movements (p. 7). The straight legs and the vertical body, neck, and head, together with immobility and extreme feather-sleeking, assist in concealment. I agree with Daanje (1950), who feels that the Bittern Stance has evolved (ritualized) from the second phase of the jump, and that the "freezing" in this phase "has the greatest selective value in relation to the given environment and its own colouration."

The stripes on the neck of a Green Heron holding the Bittern Stance blend with the overall vertical aspect of a background of reeds, *Phragmites communis*. The similarity is heightened by the extreme vertical thinness of the bird. Cowles (1930) remarks on the striking similarity of a Striated Heron in the Bittern Stance to a bed of reeds. *Butorides* is practically cosmopolitan; *Phragmites* is cosmopolitan. It is quite possible that the Bittern Stance of *Butorides* originally evolved as an adaptation to a reed or reed-like environment, even though today the different species of *Butorides* frequently occur in habitats free of reed vegetation.

The bulging eyes and the slight head movements help keep the predator in constant view. Rarely, a Green Heron in the Bittern Stance will turn very slowly on its perch in order to follow a passing predator, but only when the predator is some distance away. Cowles (1930) noted this same slow turning in an African *striatus* holding the Bittern Stance.

Inter-specific warning. Green Herons react to the warning calls of many species of birds and assume one of the alert postures or Bittern Stance, or fly. The heron flies only when the call is given suddenly, loudly, and very close, especially if the heron is previously unaware of the presence of the calling bird. Table 1 lists species to whose alarm notes Green Herons react.

FIGURE 6. Green Heron in the Bittern Stance.

TABLE 1

BIRDS AND THEIR CALLS WHICH RELEASE ALERT POSTURES AND/OR FLIGHT IN THE GREEN HERON.

Species	Call
Great White and Great Blue Herons, *Ardea herodias*	"frawnk"
Little Blue Heron, *Florida caerulea*	"aaah"
Reddish Egret, *Dichromanassa rufescens*	"crog"
Common Egret, *Casmerodius albus*	"frawnk"
Snowy Egret, *Leucophoyx thula*	"aarg"
Louisiana Heron, *Hydranassa tricolor*	"raah"
Black-crowned Night Heron, *Nycticorax nycticorax*	"squok"
Yellow-crowned Night Heron, *Nyctanassa violacea*	"squak"
American Bittern, *Botaurus lentiginosus*	"wok-wok"
Tiger Bittern, *Tigrisoma mexicanum*	"wok-wok"
Fish Crow, *Corvus ossifragus*	"caw-caw"
Redwinged Blackbird, *Agelaius phoeniceus*	"check"
Common Grackle, *Quiscalus quiscula*	"chuck"

I observed no responses to the alarm notes of a number of warbler species (Parulidae), sparrows, or the many Herring Gulls, *Larus argentatus* Pontoppidan, that flew over the breeding areas on Rulers Bar Hassock. Surprisingly, the loud, high-pitched alarm calls of the Killdeer, *Charadrius vociferus* Linnaeus, rarely evoked even a mild alert posture from a Green Heron. Many of the species listed in Table 1 bred in the same areas as the Green Herons, while the Killdeer on Rulers Bar Hassock nested around the heron breeding sites. Gulls bred in the marshes far from the heronry, so the inter-specific responses of the Green Heron may be conditioned by learning the warning calls of nearby breeding associates.

Responses to other stimuli. When they first arrive on the breeding grounds, Green Herons react, sometimes violently, to a number of sounds with no biological meaning (roar of fire engines, police sirens, and aircraft). In most cases, habituation (in Thorpe's sense, 1956) rapidly leads the birds to ignore such extraneous sounds. But one pair may become habituated while another never learns. Male and female 1-13 were under daily observation from May 9 to June 15, 1955, on Rulers Bar Hassock. A small abandoned settlement, "The Raunt," about 200 yards from nest 1-13, was set on fire by local juveniles almost every weekend during this period. I was in my blind near nest 1-13 on May 14, 1955, when the fires were set for the first time after the arrival of the Green Herons. When the fire engines came roaring north on Cross Bay Boulevard, male 1-13, which was incubating one egg at this time, fairly exploded from the nest, almost knocking the egg out. The bird flew out of sight, but returned shortly after the fire engines left. The male's flight on the following weekend was not quite as precipitous, but the bird did leave the nest hurriedly. The female flew violently from the nest later the same day after the engines were recalled to the burning "Raunt." The following weekend, male 1-13 only stood over the eggs when the fire trucks roared by, but the bird was obviously tense. Female 1-13 merely turned her head in the direction of the sounds. The male continued to rise from the eggs or young throughout the observational period, but the female finally ignored the sirens completely.

The opposite of these sexual differences was found at nest 1-12. Here, the male became habituated very rapidly, but the female never did, and her precipitous flight from the nest lasted throughout the breeding season.

Responses to predators. Green Herons respond in many ways to the approach of predators. An incubating bird invariably threatens and/or attacks an approaching Common Grackle, *Quiscalus quiscula* (Linnaeus). I have never seen a Green Heron intimidated by grackles, which only take Green Heron eggs from unguarded nests. Grackles are smaller than Green Herons and are usually intimidated by them.

These herons, however, respond to the egg-hunting of Fish Crows, *Corvus ossifragus* Wilson, in various ways. Some threaten the crows briefly and then leave, some leave their eggs as soon as the crow advances toward the nest, while a few threaten, attack, and drive away the predator. Green Herons with young will usually remain and act hostile, but action varies with the individual. The Fish Crow has a more formidable threat display than the Grackle. Even the large Black-crowned Night Herons and Yellow-crowned Night Herons, *Nyctanassa violacea* (Linnaeus), are quick to leave their eggs and young. Thus the intensity of the displays and the persistence of the crows, not their size and aspect, must be of chief importance.

Green Herons are more prone to flee and land and then scold after they are frightened by a person if young are in the nest, and they are more likely to fly far away if the nest contains eggs. Presumably the calling has survival value in that it stimulates young nestlings to crouch, and older ones to give the Bittern Stance. The "raaah" calls appear to have three functions: (1) to intimidate a potential predator; (2) to warn the young; and (3) to warn nearby herons. Sometimes adults flushed from nests containing eggs give one or more "skyow" calls as they fly away. Nearby Green Herons become alerted, actions which suggest a social warning function for the "skyow" call.

Only those Green Herons whose nests are on or very close to the ground react to the presence of a dog. They rise to a nearby branch without trying to peck at the dog and give the "skuk-skuk" or "raaah" call. The dog usually looks briefly at the heron and then trots off.

MIGRATION AND BEHAVIOR ON ARRIVAL

The Green Heron moves northward in the late winter and early spring from wintering grounds in northern South America, Central

America, Mexico, and the southern portions of the Gulf states. Large migratory flocks are mentioned in the literature (Scott, 1890); and many authors speak of nocturnal migrations (for example, Weston, cited in Howell, 1932; Cruickshank, 1942). Griscom (1923) states, "especially in August Green Herons are often heard migrating at night in extraordinary [sic] numbers."

Sexing on the basis of observed behavior and soft-part color suggests to me that both sexes arrive on the breeding grounds at the same time. Allen's (1929) statement that the females arrive first does not hold for Rulers Bar Hassock, where I made observations during the years 1954, 1955, and 1957. Table 2 lists significant events during the breeding seasons of 1955 and 1957.

TABLE 2

ARRIVAL BEHAVIOR, SIGNIFICANT NESTING EVENTS, AND DEPARTURE OF GREEN HERONS, RULERS BAR HASSOCK, SEASONS OF 1955, 1957.

Calendar of Events	Date of Occurrence			
	1955		1957	
First Green Heron arrives	April	16	April	20
First Flying Around behavior	"	20	"	21
First Pursuit Flight	"	20	"	21
First "Skowing"	"	22	"	21
First Stretch Display	"	23	"	27
First Flap Flight Display	"	23	"	22
First prolonged nest repairs	"	28	"	27
First egg of season laid	"	30	"	29
First young hatched	May	22	May	21
First brood fledged	June	13	June	11
First egg of second nesting	"	22	unknown	
Last egg of season laid	August	4	unknown	
Last Green Heron observed on area	October	6	unknown	

Migration. In each of the three years the first Green Heron arrived during the third week in April. I agree with Cruickshank's (1942) statement, "the first widespread wave generally arrives by the end of the third week in April."

Table 3 lists the time of arrival, numbers, and flight paths of Green Herons believed to be migrating, observed on Rulers Bar Hassock from April 16 to April 23, 1955. Many of these records are of two birds arriving together. Some of the migrants landed in the breeding areas and seemed to stay. Others did not land but continued northward, while a few settled for a few minutes and then resumed their northward flights. In 1955, at my request, Dr. William H. Drury, Jr., then of Cambridge, Massachusetts, made observations of the arrival of Green Herons in eastern Massachusetts. He commented that the species arrived in 1955 most commonly in twos. Figure 7 shows the

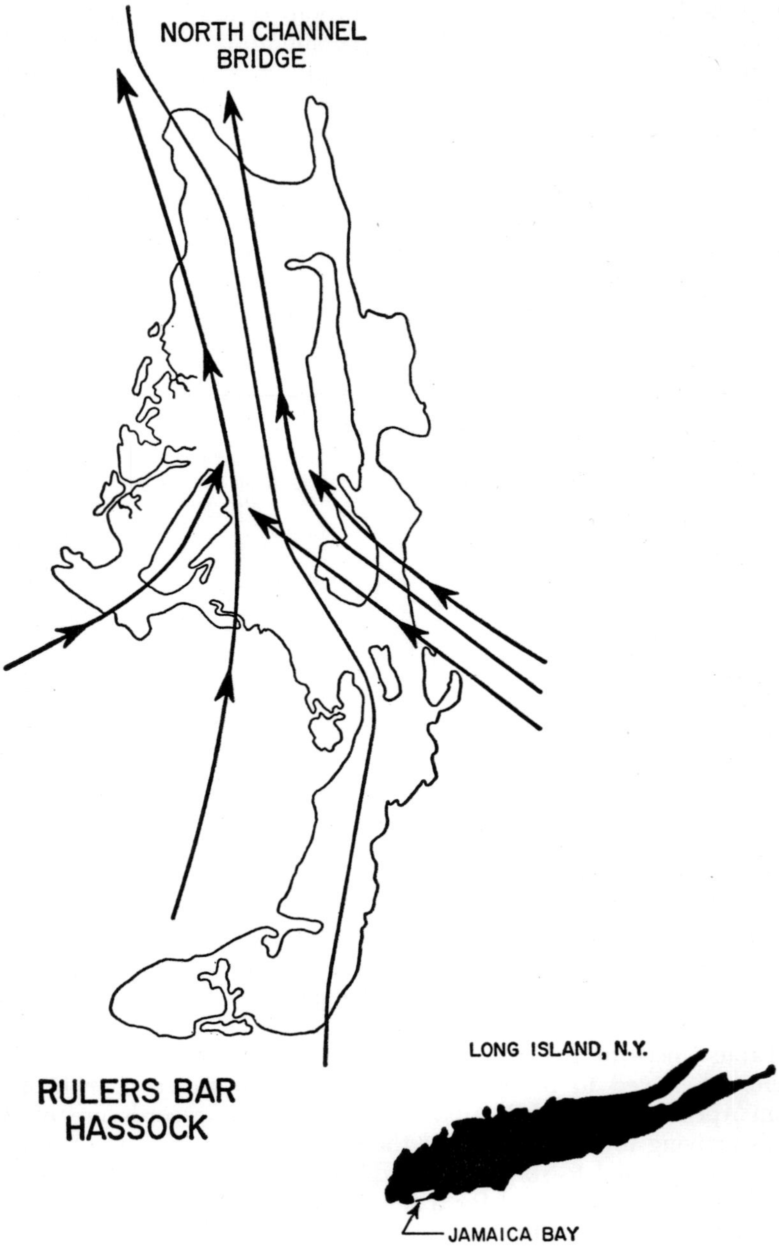

FIGURE 7. Map of Rulers Bar Hassock showing main migration routes of Green Herons.

principal lines of flight to and beyond Rulers Bar Hassock during the same period indicated in Table 3. Most herons flew at a height of about 25–50 feet. At no time did they call in flight.

TABLE 3

TIME OF ARRIVAL, NUMBERS, AND FLIGHT PATHS OF MIGRATING GREEN HERONS ON RULERS BAR HASSOCK, APRIL, 1955.*

Date	Time	Number	Flight Path
April 16	1130	1	Army FA, N to 1 BA, lands, flushes N out of sight
" "	1201	1	Raunt FA, N out of sight
April 19	1655	1	Raunt FA, over CBB, lands in 1 BA
" "	1701	1	Army FA, N to 1 BA, lands
" "	1748	2	Army FA, N to 1 BA, land 100 yards apart, roost
" "	1801	3	Army FA, N to 1 BA, land
" "	1807	2	Raunt FA, N to 4 BA, land and roost
April 20	0802	1	Army FA, N to 1 BA, lands and stays
" "	0803	1	W along dike to 1 FA, lands and feeds
" "	0805	1	Army FA, N to 1 BA, lands
" "	0941	1	Raunt FA, N to 4 BA, lands
" "	1213	1	N on CBB, speed was timed–lost sight of near NCB
" "	1748	2	Raunt FA, N to 4 BA, land
" "	1817	1	Army FA, N to 1 BA, lands
" "	1819	2	Army FA, N to 1 BA, land apart, roost
April 21	1615	1	Army FA, N to 1 BA, lands
" "	1625	3	W along dike to 1 FA, then to 1 BA, land
" "	1805	2	Army FA, N to 1 BA, land
April 22	1700	2	Army FA, N to 1 BA, land
" "	1815	3	Army FA, N to 1 BA, land
" "	1817	2	Army FA, N to 1 BA, land
" "	1818	1	Army FA, N to 1 BA, lands
" "	1819	1	Army FA, N to 1 BA, lands
" "	1829	1	Army FA, N to 1 BA, lands
April 23	1444	1	Raunt FA, N to 4 BA, lands
" "	1733	1	Raunt FA, N to 4 BA, lands
" "	1736	2	Raunt FA, N to 4 BA, land apart
" "	1737	1	Army FA, N to 1 BA, lands
" "	1802	1	Raunt FA, N to 4 BA, lands

*Abbreviations as in Figure 1. All times are EST.

I have never seen "flocks of from 20 to 50 individuals" as recorded by Audubon (1840) and quoted by Townsend (in Bent, 1926); it is possible, however, that larger flocks may have dispersed before they reached Long Island. Furthermore, during my three years of observation on Rulers Bar Hassock, I found no traces of any migratory movements at night. Migration appeared to end shortly after sunset; the birds landed and roosted in reeds and bushes. I spent several nights on Rulers Bar Hassock during the migration period but heard no migrating Green Herons. The diurnal migrants did not call; if silent, nocturnal migrants would be missed. Audubon (*op. cit.*) states that the large flocks he saw landed for the day just after sunrise; I saw no such movements at any time, and I was on the observation platform at least one hour before sunrise and remained there for at least one hour after sunset.

Many of the early migrants on Rulers Bar Hassock had brilliant soft-part colors, but because of the generally poor visibility, no general statement can be made for all migrants. Later migrants varied from brilliantly colored to dull colored individuals.

Behavior. Early migrants rarely shifted from place to place after settling in the roost. Aggressive behavior during the early arrival stage was limited to an occasional "skow" or Forward Display. Many of the early migrants appeared to be tired.

In their close attachment to their territorial site after arrival, Green Herons differ from the other herons on Rulers Bar Hassock. The former move in at once and start reproductive activities. I saw no dancing ground behavior by Green Herons before occupying the nesting sites.

HOSTILE AND SEXUAL BEHAVIOR

I use "hostile behavior" for activities that appear to express attack motivation and "sexual behavior" for displays that have few or no apparent hostile or other non-sexual components. However, the descriptions that follow show that the designation "hostile" *or* "sexual" is frequently artificial. It is difficult to interpret the range of expression resulting from interactions of varying drive strengths as attack, escape, sexual, or otherwise. Recent ethological publications emphasize the roles of the attack and escape drives in the causation of threat displays (see, for example, the review by Moynihan, 1955a), but the relation of these two drives with others, principally the sex drive, must be considered (Hinde, 1953a).

For example, crest-raising is a component of many Green Heron displays. In the Full Forward Display it is a hostile component, but in the extended alert posture it may express conflict (p. 15). Also, crest-raising in some displays seems to express general excitedness, separate from hostility or sexuality.

I should like to offer an example as an argument against pigeonholing displays as sexual *or* hostile. Moynihan (1955b) states, "they [i.e., appeasement displays] are produced, of course, by the usual type of hostile motivation, but the escape drive of an appeasing bird is almost always stronger, usually much stronger, than the attack drive." The Stretch Display of the male Green Heron is an appeasement display *in certain circumstances.* Although male Green Herons usually orient their Stretch Displays toward a female, they frequently perform Stretch Displays when no other Green Herons are in sight. I cannot imagine the activation of a strong escape drive in such males. Moynihan (*op. cit.*), however, does add, "the immediate causation of many appeasement displays, apparently including three or more distinctly different drives, is far more complicated than that of the majority of threat displays."

Moynihan assigns a subsidiary role in pair formation to advertising songs, their primary role being intimidatory in nature. My work on the Green Heron shows that at an early stage, the male's "skow" functions not only as a long distance threat, but also as a "sex recognition marker" for unmated females. Later, the male's advertising calls stimulate one particular female to return to the vicinity of the male's display post and engage in mutual sexual activities. One function of the female's "skeow" appears to be to stimulate the male to perform sexual displays. These problems will be discussed later, but the assignment of *only* a threat or intimidatory function to the advertising calls of the Green Heron would be misleading.

Many displays, at least those in the birds with which I am most familiar, grade from purely hostile to purely sexual. The causation of the extreme hostile and sexual displays seems clear, but for many displays of lower intensity I cannot decide whether a certain component appears to express hostile, sexual, or other motivation; furthermore, I find it difficult to apply the concept of a balance or conflict of drives in some low intensity displays.

Even though I consider them artificial categories, I shall use the terms "hostile behavior" and "sexual behavior," but *for purposes of presentation only.*

HOSTILE BEHAVIOR

All observations involving hostile behavior were of encounters among free-living birds during the breeding season. Green Herons are solitary once they have arrived on their wintering grounds, which may be more than a thousand miles from their breeding areas. At no time during the winter of 1955–56 did I see more than one Green Heron in any one area in Florida Bay. Lack of experimental data and the rarity of encounters among non-breeding birds make the evaluation of breeding season fighting difficult because of the influence of sexual and other drives.

The terms "drive" and "motivation" have been used variously in ethological and psychological literature, and in some recent ethological papers (Moynihan, 1955c; Dilger, 1956), the terms are used interchangeably. I shall use the term "tendency," following Hinde (1956a), Marler (1956), and Andrew (1956–57). For example, when it is my impression that a given individual is behaving in a sexual or hostile fashion, I shall use "sexual tendency" or "attack tendency." For a critical review of the use of the term "drive" in ethological literature, see Hinde (1956a).

Because no experiments were made, the estimates of relative and actual strengths of the tendencies I have assigned to some displays must be considered strictly descriptive.

I recognize the following four overlapping types of hostile behavior: (1) non-aerial, non-vocal; (2) non-aerial, vocal; (3) aerial, non-vocal; and (4) aerial, vocal. This division is for presentation only, and does not imply that this is the only method of classification.

Non-aerial, Non-vocal Displays

Crest-raising. The crest of the Green Heron is the most mobile of its feather units, and there is great variation in crest-elevation (Plate 7). Green Herons show crest-raising in numerous situations. Moderate crest-raising, together with alternating erection and depression, is characteristic of conflict situations, while extreme erection is always an expression of high intensity attack tendencies. Full erection always accompanies the Full Forward Display, and some degree of erection is shown by a bird assuming the Forward Display. A male on its territory, surprised by the sudden appearance of another Green Heron, shows crest up-down behavior, but if the intruder alights on or close to the territory, the owner assumes the forward posture with crest erect. The alternating crest up-down may indicate a conflict of escape-attack tendencies.

Green Herons landing anywhere on the breeding area, including their own territory, always erect their crest on landing. It seems as if territorial stimuli trigger such crest-raising; a bird landing in its own territory is presumably prepared to defend the territory, while an intruder is prepared to defend itself against the attacks of the owner, even if the owner is not present when the intruder alights. Green Herons landing out in the vast feeding areas surrounding their breeding sites at Rulers Bar Hassock rarely erected their crest on landing. Exceptions occurred in the proximity of other feeding Green Herons. At Newburyport Harbor, Massachusetts, where Green Herons defended feeding areas, crest-erection on landing in the feeding areas was usual.

Habituation and recognition of neighbors may account for peculiar changes in hostile behavior exhibited at later stages in the breeding season. As the male's behavior shifts from hostile to sexual displays, the territory shrinks until only the nest and a few feet around it are defended. A bird defending such a small territory rarely shows any crest-erection at the approach of a neighbor. Recent arrivals, however, that have just been able to carve out a territory, apparently because of the shift from hostile to sexual behavior by the earlier arrivals, show vigorous hostile behavior toward these neighbors, but only for a brief period. As soon as the late arrivals direct their displays toward one female, the flights of nesting neighbors are almost completely ignored. Crest-raising is almost completely eliminated among neighbors at the same stage of the breeding cycle. The decrease of hostile responses, however, may be either the result of over-riding parental tendencies or of learning the characteristics of neighbors.

When adults are disturbed at the nest, crest-raising from the Forward Display posture is typical if young are present. If eggs are in the nest, the flushed bird rapidly raises and lowers its crest.

At times, slight crest-raising appears to be a low intensity intention movement of attack. The full erection accompanying the Full Forward Display makes the displaying bird appear much larger. The crest is dark, glossy, greenish-black, sharply set off from the maroon of the neck. This contrast emphasizes crest-raising. The color and striking movements of the crest suggest a true releasing function. Presumably the primary function is to increase the apparent size of the bird, thus helping to intimidate an opponent; that is, to release avoidance in the latter. However, the objects of the display may react otherwise. Some respond by displaying hostility in return; some react with a direct attack; while some ignore the display. Crest-raising may also have a secondary function as a social warning. Birds flushed are made much more conspicuous by alternately raising and lowering the crest, thereby alerting their neighbors.

I doubt that crest-raising in any alert posture indicates an activated attack tendency (p. 15). Moynihan (1955b), however, feels that *both* attack and escape drives are activated simultaneously; that is, whenever one or the other of these two drives is activated, the other is also activated. Dilger's (1956) work on the North American thrushes reveals that crest-raising in these birds is usually associated with an increased attack motivation, but Marler (1956) associates crest-raising in the Chaffinch, *Fringilla coelebs* Linnaeus, with escape motivation.

A Green Heron fleeing from the direct attack of another Green Heron does not erect its crest during the escape flight. During a Pursuit Flight the pursued often erects its crest as the pursuer draws near. Shortly after alighting, the pursued individual assumes the normal perch posture (Plate 6) with no crest-erection if the attacker does not continue or renew the attack.

The crest is never raised in pure escape movements such as the Bittern Stance or the Withdrawn Crouch (p. 36). On one occasion I saw a Green Heron in the Bittern Stance erect its crest as it very slowly turned its head to watch a Marsh Hawk, but only after the hawk had flown some distance from the heron.

The causal analysis of crest-raising is further complicated by its occurrence in the Crooked Neck Flight Display (p. 41). In this display a raised crest emphasizes what is primarily a sexual display. I have dwelt at some length on crest-raising because it is a component of so many diverse displays, and its meaning varies with each display.

Tail-flipping. The degree and intensity of tail-flipping vary with the intensity of the tendencies involved. A bird attacking an opponent

FIGURE 8. Green Heron in the Forward Display. Dotted lines indicate tail-flipping.

from the Full Forward Display posture always shows exaggerated tail-flipping, both in intensity and frequency. The tail is moved, as far as I was able to tell, only in a vertical plane. A bird in one of the alert postures frequently shows slight tail-flipping, a flight preparation movement; in hostile displays flipping becomes greatly exaggerated.

Feeding Green Herons frequently flip their tails just before striking, but the causal basis here seems obscure. Tail-flipping, like crest-raising, may also serve as a social warning signal because alarmed birds are thus made more conspicuous. Birds in the Bittern Stance or the Withdrawn Crouch never flip their tails.

Forward Display. A bird assuming this display posture (Figure 8) from the normal perch posture (Plate 6) moves as follows:

1. lowers the head and body to the horizontal
2. withdraws the head and neck so as to appear neckless
3. bends its legs at the heel joint, thus assuming a crouched position
4. erects its crest and keeps the feathers erect throughout the display
5. erects neck, breast, back, and flank feathers moderately, and erects back (scapular) plumes slightly
6. if it flips its tail, the motion is slow
7. holds its bill in a horizontal position and keeps it closed
8. orients its body to face the opponent.

The Forward Display almost always precedes the Full Forward Display. In one variant of the Forward that I have called "stalking," the displaying bird assumes the Forward Display, then walks slowly and

FIGURE 9. Green Heron in the Full Forward Display. Dotted lines indicate extreme tail-flipping.

deliberately toward the opponent. Males displaying on the nest platform frequently alternate Forwards with stalking around the nest rim.

Full Forward Display. Except for actual attack, this display expresses the highest level of the attack tendency (Figure 9). From the Forward posture, the bird moves as follows:

1. extends the head and neck fully forward, body still horizontal
2. erects crest, neck, back, breast, and flank feathers to an extreme, and keeps them erected throughout the display
3. erects scapular plumes in a fan-like manner
4. flips tail vigorously
5. bulges eyes (iris may change from yellow to deep orange)
6. opens its bill, showing the rich, red lining of its mouth
7. lunges toward the opponent, extends its wings slightly and waves them
8. at the height of the lunge, utters rasping "raaah" calls (p. 38).

The bulging, yellow-orange eyes of the displaying bird are accentuated by the dark, blue-black lores, and the effect of the gaping bill is heightened by the rich, red lining of the mouth. Of all the threat displays of the Green Heron, the Full Forward is the most effective because the opponent is almost always intimidated, retreating immediately. If, however, the opponent displays in return, a vicious fight ensues.

An aerial variant of the Full Forward occurs when the attacker flies a short distance toward a displaying opponent.

Fighting. Two Green Herons engaged in fighting are so preoccupied that I was able on several occasions to approach and almost touch fighting birds. During contact both combatants peck viciously at each other, mostly at each other's head. They flail away with extended wings trying to buffet each other. It is characteristic of fighting Green Herons that they stab from a horizontal stance, rather than from the upright position of many other herons.

Fights are rare and usually last less than a minute or two. I have never seen a fight that appeared to result in real bodily harm. The fight usually ends when one bird flees giving several high-pitched "skeow" calls.

Stiff-necked Upright Display. This is the rarest of the threat displays, and it is difficult to analyze. It is my impression that in some situations the strengths of three tendencies—attack, escape, and sexual—are all high and in conflict; in other situations only attack and escape tendencies are intense and in conflict; while occasionally attack and sexual tendencies are involved.

I have seen the Stiff-necked Upright Display in males just before and immediately following an attempted copulation with a non-receptive female. These males had just been vigorously displaying the Stretch and/or Snap Displays (p. 33). As the female approached the male on his territory, the latter showed the Stiff-necked Upright Display instead of the normally expected Forward or Full Forward Display. The male suddenly stretched his head and neck upward and slightly forward, erected the neck feathers, giving the neck a very thick, stiff look, and erected the crest slightly. Crest feather movements were variable. The aspect of the male was tense and rigid, and he moved in a very stiff, awkward manner toward the female (Figure 10).

If the female suddenly moved toward the male, the latter pecked down at her, but the pecks were short, and the male immediately retracted his head and neck, as if the thrusts were being delivered in a half-hearted manner. The appearance of the male was that of a bird that wanted to attack, flee, and copulate all at the same time. If the female turned from the male, the latter attempted to mount her, but never successfully as far as I could see.

Always after the thwarted copulatory attempt, the male showed what appeared to be displacement preening from the stiff-necked position. In normal preening of the neck region the bird stretches its head and neck up and slightly forward, but there is little or no erection of crest feathers, and the preening is accomplished with deliberate, smooth strokes. The neck-preening shown from the stiff-necked position is rapid and performed with jerky, uneven strokes; it is never complete. This display is never accompanied by a call.

FIGURE 10. Green Heron in the Stiff-necked Upright Display.

Possibly the Stiff-necked Upright Display has been ritualized from displacement neck preening with hostile elements superimposed. The appearance of preening movements after an unsuccessful copulatory attempt may indicate thwarting of the sexual tendency. The actual bodily contact resulting from the abortive copulatory attempt may serve to increase the sexual tendency. An alternative explanation for the incomplete neck-preening is that the bird is simply smoothing feathers that have been disarranged during the display. The continued presence of the female may account for the brevity of the male's preening movements.

The Stiff-necked Upright Display is also shown under conditions that may express antagonistic activation of attack and escape tendencies. When "skowing," typically a hostile display (p. 36), a territorial male normally adopts the Forward Display if another individual lands on its defended area. Under certain conditions, however, the defending male will assume the Stiff-necked Upright Display. If the intruder is a male and "skows" upon landing, the territory holder assumes the Stiff-necked Upright. The "skowing" and the sudden appearance of the intruder might momentarily activate a strong escape tendency in the territory holder. This display is shown only for a few moments, and in almost every case the intruder fled when the owner displayed. In exceptional cases the intruder did not flee until the owner assumed the Full Forward Display.

On one occasion I saw the Stiff-necked Upright Display when it seemed to involve both attack and sexual tendencies. A recently formed pair were showing mutual feather-nibbling, normally a prelude to copulation, when suddenly a neighboring male landed in the same tree and "skowed" once. The male of the pair immediately assumed the Stiff-necked Upright Display and lashed out and down at the intruder, which immediately fled. It seems likely that in this situation, the displaying male's sexual tendency had been quite intense, and that the sudden appearance of the intruding male activated a strong attack tendency, resulting in the Stiff-necked Upright Display.

Bill-snapping. This action is shown before copulation, especially before the first few copulations. The activity involves an interaction of hostile and sexual elements. Both sexes show this behavior, but the male shows more hostile elements in bill-snapping than the female, although they are of low intensity. As the male approaches the female, he adopts the forward position with very slight erection of crest and neck feathers; he extends his head and neck forward, then rapidly clicks his mandibles together close to the head of the female. She reacts by adopting the forward position, extending her head and neck toward the male's head, and then she bill-snaps. The crest and neck feathers of the female are barely erected.

Feather-nibbling. Both birds usually engage in mutual feather-

FIGURE 11. The Snap Display or *Schnappbewegung* of the Green Heron.

nibbling before copulation. The head, neck, and flank feathers of the partner are gently grasped in the tips of the mandibles, and then the feathers are rapidly released and grasped again repeatedly. Both sexes may flutter their wings during feather-nibbling activities.

A sudden movement on the part of either bird causes its partner to retreat at once, showing a low intensity Forward Display. A few moments later the bird that has retreated again approaches its partner, showing bill-snapping, then feather-nibbling behavior. Bill-snapping and feather-nibbling are reduced to a minimum after the first few copulations.

Snap Display. A male when snapping assumes the forward position, extends his head and neck fully forward and down, and then snaps his mandibles together to produce an audible click. The crest and neck feathers are slightly erected (Figure 11). I believe that these feather movements are indicative of low intensity hostile tendencies, because a male approached by a female during or immediately after the performance of a Snap Display may show more intense hostile behavior toward the female.

This display is the *Schnappbewegung* of the Common Heron, *Ardea cinerea* Linnaeus (Verwey, 1930). The movements of *B. virescens* are very similar to those of *A. cinerea,* but it is my impression that the snap of *cinerea* has been more ritualized than the snap of the Green

Heron, because additional, highly exaggerated movements are involved in the Snap Display of *virescens*. The movements of *cinerea* are few and smoothly performed, indicating, according to the "rules," more advanced ritualization. Daanje (1950) and Tinbergen (1954) have indicated that the course of ritualization involves exaggeration and simplification of movements.

I have never seen female Green Herons perform the Snap Display, and neither has Verwey in *cinerea*; de Waard (1937)[1] and Baerends and Baerends (1950:152) claim otherwise for *cinerea*.

I regard the Snap Display of the Green Heron primarily as a sexual display, but it appears to contain some hostile elements.

A sexually active male Green Heron will show a rapid alternation of Stretch and Snap Displays, and the longer the male's sexual tendencies remain unsatisfied, the more frequent and the more intense his Snap Displays will be. After prolonged and intense sexual displaying, the male frequently shows modified versions of the Snap Display. For example, the bird may rarely erect its feathers, and it adds bobbing and bowing movements. With increased sexual activity, the whole body of the displaying male vibrates, and the branch on which the male is perched shakes violently with the intense bobbing motions. A male may begin a bout of Snap Displays by first extending the head and neck out horizontally, then bowing low very slowly; the initial bow may then be followed by a fantastic series of intense bobs, then a few slow bows, and then a repetition of the entire sequence. Males showing such intense Snap Displays usually shift readily to bill-snapping and feather-nibbling at the approach of a female. This shift probably indicates a sharp reduction in hostility on the part of the performing male. Sometimes a male showing such intense Snap Displays over a long period of time will attempt to force copulation on an approaching female, and under these conditions, the Stiff-necked Upright Display described previously may be shown.

Verwey (*op. cit.*) and Tinbergen (1952) conclude that the Snap Display of *cinerea* has been derived from the fish-catching movements, but I feel that the display is more likely ritualized twig-grasping, a component of nest-building behavior. Verwey saw the movement in great detail only in *cinerea*, where from his description and fine drawing (adapted from Holstein, 1927), the action certainly resembles fish-catching movements. My comparison of snap movements in *virescens* and other North American heron species, including the Great White and Great Blue Herons, points to a derivation from nest-building movements. In typical nest-building, the male Green Heron secures

[1] I have not seen de Waard's paper. Baerends and Baerends state that their own unpublished data reveal only occasional snap movements in female *cinerea*. My own data include no observations of a female showing the snap; unmarked females, however, might be mistaken for males.

FIGURE 12. The Withdrawn Crouch escape posture of the Green Heron.

a twig and passes it to the female standing on the nest platform; the latter inserts the twig into the nest with the highly characteristic "tremble-shoving" movements (Lorenz, 1955). When grasping a twig, the male does so by reaching out and down, head and neck fully extended forward, and grasping the twig between the tips of the mandibles. During functional twig-grasping, the crest, neck and flank feathers are never erected; in addition, the movements are slow and smooth, not formalized as in the Snap Display. Males showing the Snap Display occasionally seize, then release, a twig during the performance. Rarely, a twig will be actually broken off, but it is always dropped at once. Verwey stated that *cinerea* never seized anything during the Snap Display, but my observations of *herodias* show that twigs may be grasped momentarily during the snap performance.

It might be argued that low intensity nest-building tendencies are shown by a male performing the Snap Display, but in the derived twig-grasping movements, as described above, such movements only partly resemble typical or functional twig-grasping. The incomplete twig-grasping seems to favor the derivation of such movements in the Snap Display from nest-building activities, because Green Herons showing true, but low intensity, twig-grasping normally seize and wrestle with the twig for some time before dropping it. They may move about the tree, inspecting and grasping a number of twigs. If functional twig-grasping behavior was being shown in the Snap Display, we would expect an occasional male to break off a twig and carry it to the nest, but I have never seen such behavior prior to pair formation. After the pair is formed, the male no longer engages in Snap Displays, and his twig-grasping at this time is definitely functional.

Withdrawn Crouch. This posture indicates high escape tendencies, and it is assumed by a Green Heron when surprised by the sudden appearance of another bird, typically another Green Heron, from above. The head is completely withdrawn, the plumage is flattened, and the bird crouches low (Figure 12). A bird may change from the Withdrawn Crouch to a number of other postures depending upon the nature of the disturbance. For example, if a passing Redwinged Blackbird, *Agelaius phoeniceus* (Linnaeus), or Common Grackle has evoked this response, the posturing heron normally reverts to the normal perch posture in a matter of seconds, but if the intruder is another Green Heron, the bird in the Withdrawn Crouch may flee or assume the Forward Display, depending on circumstances.

NON-AERIAL, VOCAL DISPLAYS

Advertising calls. Males and females advertise their presence in the breeding areas with distinct calls. During the first few days after arrival on the breeding area, both sexes may call anywhere, but after a few days, the males especially restrict their calling to more and more limited areas. The male takes a position in the top of a tree or other exposed perch, raises his head and bill to about a 45 degree angle and gives a single, low-pitched "skow" (Figure 13). The male frequently stands on the floor of an old nest and "skows" from this position. If it has been selected as the future nest site, the male concentrates his "skowing" on this nest or its immediate vicinity, and defends the nest vigorously.

The first adequate description of the "skow" call of male Green Herons is that of Simmons (1925), near Austin, Texas. Simmons describes the call as "skah-ooow (long o)" and states that it may be heard from a distance of about seventy feet. Townsend (1928) described the "song" in much greater detail, stating that the calls sounded like deep groans, "ow-er," the *o* as in cow, *er* as in the German *ch* [sic]; hence, "ow-ch." The typical call of the males breeding on Rulers Bar Hassock was of the one note variety, that is, a single, guttural "skow." A few individual males gave calls of much lower pitch that can best be described as "skow-ch," resembling Townsend's "ow-ch."

My observations of Green Herons in widely separated localities supply interesting data on geographical variation in the pitch of the "skow" call. In Rulers Bar Hassock, most of the males uttered a moderately low-pitched "skow," some gave a very low-pitched "skow-ch," and a few others gave a high-pitched "skow." In eastern Massachusetts (Wayland, Burlington), all males gave the moderately low-pitched "skow." In Florida Bay, all males gave the highest pitched

FIGURE 13. Male Green Heron "skowing."

"skow" of any Green Herons I have observed. On the other hand, males around the Laguna de la Joyas near Puerto Arista, Chiapas, Mexico, uttered the lowest-pitched "skow" calls I have ever heard; these males occasionally uttered the two-parted "skow-ch," still very low-pitched. Simmons' description, that is, a long, two-parted "skah-ooow," may indicate another variation of the basic call. Fisher's (1933) rendition of the call given by a startled Green Heron, "scow-scow," is probably his translation of the typical "skeow" or "skyow" call described on p. 13, and is not to be considered the advertising call.

The "skow" of Green Herons is certainly analogous, if not homologous, to the "rwo" call of *A. cinerea* (Verwey). Both the "skow" and "rwo" calls closely resemble calls given outside the breeding season, and both seem to function as a long distance threat and as an attraction call for females. Verwey, however, calls the "rwo" of *cinerea* a mating call, because the calling male stops if a female is attracted, whereas the male continues to call if the reacting individual turns out to be a male. Verwey feels that the biological significance of the "rwo" call is attraction of the female from some distance. However, the fact that male Common Herons renew their "rwo" calling if another male calls points to a threat function. This is certainly the case with Green Herons.

The advertising call of female Green Herons resembles the alarm call of both sexes (p. 13). The female may or may not raise her head and bill when calling, while the male always erects his head to some degree. The advertising call of the female is a single, rather high-pitched "skeow" or "skyow." A female attracted by the "skowing" of a male usually perches in an exposed position near the male, and a "skowing-skeowing" bout normally follows. I could find no evidence of any geographical variation in the "skeow" call of female Green Herons. Verwey states that the female Common Heron does not have the mating call.

Attack calls. Males or females attacking an opponent from the Full Forward Display almost always give a series of low, harsh "raaah" calls. Both sexes use the same call when threatening a potential egg predator such as a Common Grackle or Fish Crow. I have heard Green Herons on their territories use the same "raaah" call from the Full Forward position in threatening Black-crowned Night Herons. The display of the Green Heron intimidated the Black-crowns in every case. Normally, the much larger, bulkier Black-crowns force Green Herons to leave a perch.

Escape and Alarm calls. These calls have been described on p. 13.

Aerial, Non-vocal Displays

Flying Around. Lone birds were observed flying back and forth over one breeding area or from one breeding area to another early in the breeding season. The flights were interspersed with preening on some exposed perch. Often the birds ended the displays by flying directly to one of the surrounding feeding areas. Many times the herons did not feed. They landed for a minute or two and then flew back. There is no tie at this point to one particular area. The details of flight during this behavior are identical with those of normal flight, but the orientation is limited to the breeding areas. Crest-raising is exceedingly common. This behavior appears to be hostile because a bird Flying Around rapidly shifts into higher intensity hostile patterns such as Pursuit Flights and Supplanting Attacks. When the breeding areas are first occupied, Flying Around is followed occasionally by pursuit of another Green Heron on the breeding area, but shortly, within a day or two, birds Flying Around readily pursue and make Supplanting Attacks throughout the day. Birds that perch after a bout of Flying Around respond readily to another Green Heron, and they will interrupt their activities to fly toward the intruder, even if several hundred yards distant. Perched individuals will assume a low intensity alert posture as the first indication of their awareness of an approaching Green Heron, and they frequently show crest-raising. "Skeows" (warning call of both sexes) anywhere within hearing range release "alerts" immediately. The function(s) of this behavior seems to be to familiarize the newly arrived birds with the habitat, with the available old nests (exposed early in the breeding season), and with other Green Herons.

Pursuit Flight. Aerial hostilities are conspicuous during the early part of the breeding period. A Pursuit Flight typically involves two birds, but I have observed on many occasions three to five birds involved. The flights are normally brief (50–75 yards), but long flights (over 100 yards) are not rare. Although no birds were individually marked, I am convinced that the pursuer was usually a male, and the pursued either a male or female. Occasionally, pursuing males uttered the "skow" call in flight, and at times the pursued gave the "skeow" call in flight (p. 43). "Skowing" males were more aggressive and persistent in their Pursuit Flights than non-calling males; that is, a calling male often changed to more intense hostile behavior (for example, Supplanting Attacks or Forward Displays on landing), whereas non-calling males usually broke off the pursuit and returned to another part of the breeding area. I rarely observed a pursuer overtake and attempt to attack a bird in flight. When this happened, the attack consisted of several attempts to nip the tail of the pursued. The pursuer erected its crest every time it made a nip. If the pursuer drew very close, it also erected its neck feathers. The pursued frequently erected its crest feathers briefly, or very rapidly raised and lowered them. The pursued characteristically flies with its head and neck raised, which enables it to watch the pursuer closely. Occasionally the pursued turned its head slightly to keep the pursuer in view.

I believe that most Pursuit Flights are expressions of rather low attack (pursuer) and escape (pursued) tendencies. Additional evidence exists in the unhurried nature and shortness of the majority of Pursuit Flights. Furthermore, the pursuer continued flying in most Pursuit Flights even if it had overtaken the lead bird, or if the pursued had landed.

Supplanting Attack. In this behavior the attacking bird flies toward a perched individual which vacates its perch. While landing, the attacking bird erects its crest and neck feathers and frequently attempts to peck at the departing bird. Supplanting Attacks are most common during the early part of the reproductive cycle, and I am convinced that the supplanting bird may be a sexually active male, because in some attacks the supplanting bird made a crude copulatory attempt on landing. In addition, a supplanted female would frequently return after a brief flight, especially if the male gave the Stretch Display. The return of the supplanted female usually triggered a bout of Supplanting Attacks and Pursuit Flights, revealing the complicated interaction of hostile and sexual tendencies. This interaction is also shown by males that gave a Flap Flight Display just before a Supplanting Attack.

Soon after the birds arrive and start Flying Around and Pursuit Flights, the first Supplanting Attacks are seen. At first, such attacks may occur anywhere on the breeding area, but in a few days these attacks are restricted to the territory defended by the male. As the territory decreases in size, the male restricts Supplanting Attacks to one female, that is, the one that continues to return even though repeatedly driven from the male's territory. A territorial male will always supplant any Green Heron it finds on its territory upon return from a feeding area, Pursuit Flight, or other trip.

Circle Flight Display. The displaying bird of either sex launches into normal flight that carries it in a rough circle back to the starting place. The diameter of the flight circle rarely exceeds fifty yards, and shorter flights of 20–25 yards are most common. I found no differences in the form of the flight from normal flight, but a displaying bird flies in a rough circle, calls frequently in flight ("skow" or "skeow"), and is usually a territorial male, or a female associating with a particular male. A male in Circle Flight Displays returns to his territory unless he has been attacked during the flight. Late in the season the female usually returns to the vicinity of the male she is interested in, but early in the breeding season she may be interrupted and wander off the Circle Flight to the territories of neighboring males. After they have made several mutual Circle Flights, however, the pair tend to restrict their displays to each other.

After a male and a female have engaged for a few days in a number of displays (Pursuit Flights, Supplanting Attacks), they may perform the Circle Flight together, and these displaying pairs become conspicuous just before the appearance of the next higher intensity form, the Crooked Neck Flight Display.

The Circle Flight differs from Flying Around in that the flight is in a rough circle, has the same starting and ending points, is rarely interrupted, and is oriented toward one individual. Pursuit Flights differ in that they may be protracted, wandering, and contain more hostile elements. Supplanting Attacks are normally performed by one bird (usually a male) directed at another individual (perched male or female). A bird giving the Circle Flight (male or female) flies around or even directly over the "object" of the flight (perched male or female), and attacks are rare.

The Circle Flight Display has more sexual than hostile components. The bird rarely raises its crest or other feathers, it rarely ends the display with a Supplanting Attack, and it orients toward a familiar individual. In addition, a female that displays the Circle Flight returns again and again to the territory of the same male, even though she is attacked repeatedly. The Circle Flight of the female definitely stimulates a territory holding male to perform a Circle Flight in return, to engage in mutual Circle Flights, or even to perform sexual displays such as the Snap Display and Flap Flight Display. The female is stimulated by these performances of the male to approach the nest tree more closely and to engage in mutual Circle Flights.

However, "skowing" and "skeowing" and occasional Supplanting Attacks and/or Pursuit Flights, point to some hostility in the Circle Flight Display. It is, however, best considered a low intensity sexual display whose function is mutual sexual stimulation to synchronize the behavior of the forming pair.

Crooked Neck Flight Display. This display is intermediate in form between the Circle Flight Display and the Flap Flight Display, and it is shown by both sexes. The flight is generally in a circle, although considerably shorter than the Circle Flight Display. Occasionally the display may involve a short, straight flight, then a return display flight; the bird may perch after the direct flight. In this display, flaps are slower and much deeper than in normal flight; this results in an undulating flight. The bird does not retract its neck fully, and it carries its neck crooked or kinked throughout the brief flight. The crest is slightly raised, and the legs may be dangled occasionally. "Skowing" and "skeowing" are common. This display appears after the pair have mutually performed the Circle Flight Display, and have been associated for some time on the male's territory.

Sexual components are more pronounced than hostile ones in this display. The Supplanting Attack that rarely ends a Crooked Neck Flight Display is half-hearted, and the supplanted bird, usually the female, shifts its perch only a few yards, then resumes its former position almost at once. This display stimulates the prospective partner to closer association and synchronization of behavior which leads to the first copulation.

Flap Flight Display. This bizzare aerial performance is highly ritualized, and it is shown more frequently by the males. The Flap Flight Display is the highest intensity form of the sequence: Circle Flight Display → Crooked Neck Flight Display → Flap Flight Display. The last display amplifies components found in the other two.

The displaying male leave its perch, often the nest platform, and makes an exaggerated flight toward a female. The wing beats are very slow and deep, and the tips appear to touch beneath the body. Lurching through the air is the best description of the slow flight. The wings produce a hollow "whoom-whoom-whoom." The bird holds its head and neck even more crooked than in the Crooked Neck Flight Display, and it usually dangles its legs throughout the flight. Males frequently "skow" and females "skeow" in flight. The flight is usually short, direct, and only occasionally in a complete circle. The male erects his

FIGURE 14. Male Green Heron showing the Flap Flight Display.

crest and neck feathers and frequently his scapular plumes to a full fan. Females erect these three feather units less, scapular plume erection especially being quite rare. On several occasions the eyes of the males bulged at the end of this display flight. Only those birds of both sexes that showed brilliant soft-part coloration engaged in this bizarre display, and the dark lores, bright orange, bulging irises, lustrous plumage, and brilliant, coral-orange legs, together with the exaggerated movements, made the whole performance striking (Figure 14).

One variant includes stalking around just before flight, and this is shown only by those males whose territories have been considerably constricted as a result of territorial encroachment by newly arrived males (p. 62). Stalking around in these cases is definitely hostile because it appears only when the female actually lands in the nest tree. Invasion of his "must defend" area stimulates the male momentarily to stalk around the nest rim, but sexual tendencies prevail, and the male then launches into a Flap Flight.

Another variant involves a series of soft "roo-roo-roo" calls just before landing after a Flap Flight. These sound like the end of the "aaroo" call in the Stretch Display. The Flap Flight with "roo-roo" calls is typically followed by a series of Stretch Displays from the nest, and I believe that such males are showing a very strong sexual tendency.

The "roo-roo" variant of the Green Heron's Flap Flight may be the first indication or the last remnant of the evolution of an aerial Stretch Display in *virescens*; such aerial Stretch Displays are common in the Reddish and Snowy Egrets.

The calls, the feather erection, and the bulging eyes may be minor hostile components, but in this flight display they emphasize and exaggerate the primarily sexual performance. The Flap Flight is preceded and/or followed by intense bouts of sexual displays such as the Stretch Display even more than is the Crooked Neck Flight Display. The supplanting "attack" that rarely ends this display is of very low intensity and suggests not a true Supplanting Attack but actually the end of the Flap Flight. The attack components of the female's display are even less; the crest, neck, and scapular feathers are barely erected, and she does not end with a Supplanting Attack.

This display, especially in the male, apparently occupies all the bird's attention. On several occasions displaying males have flown directly toward me even though I was not concealed. They landed within a few feet of me, peered for several seconds, slowly moved to an alert posture, then exploded into flight and uttered high-pitched "skeows" as they left. The displaying birds appeared ecstatic during and immediately after the display flight, and on many occasions they crashed after seeming to lose their ability to coördinate. The frequency of poor landings suggest that they are a part of the Flap Flight Display.

Only males and females that have associated closely at the male's nest tree perform this display. The function is probably to stimulate the partner sexually; that is, the start of a Flap Flight usually stimulates the perched partner to launch into the same display. Evidence for this appears in the section on Late Pairing Behavior.

Aerial, Vocal Displays

"Skowing" and "skeowing." These calls in flight have been interpreted as hostile in nature, at least for some displays (p. 39). Such calls may stimulate the partner to perform either sexual or hostile displays. The same individual may react to the same vocal stimulus in a hostile or sexual way at different stages in the breeding cycle.

SEXUAL BEHAVIOR

Only those displays preponderantly sexual, having only minor hostile components, will be described in this section.

Stretch Display. The Stretch Display of the Green Heron is the *Reckbewegung* of the Common Heron (Verwey). Because Verwey's terminology is so descriptive, I shall use his terms for describing similar displays by the Green Heron.

A male showing a typical Stretch Display performs the following

FIGURE 15. The Stretch Display or *Reckbewegung* of the male Green Heron as seen from the side.

movements from any perched position, but usually while standing on the floor of an old nest:

1. the bird stretches its head and neck upwards so that the bill, head, and neck are in a straight line pointing to the zenith, then lowers its head so that the occipital region almost touches the back, the upward and backward movements being performed in *one* smooth, uninterrupted motion
2. the bird erects its interscapular plumes to their fullest, forming a fan behind the head, during the downward movement
3. the bird now smoothly sways pendulum-like from side to side, but only the bill, head, and neck move to any appreciable extent
4. the male emits one or more very low, soft, somewhat gurgling calls best transcribed as "aaroo," "aaroo," during the swaying movements
5. the legs and feet are stationary, or moved slightly by the swaying movements
6. the crest, breast, and flank feathers are depressed
7. rarely the wings are drooped very slightly outward and downward
8. occasionally the eyes may bulge slightly, and very rarely, the iris may change from yellow or yellow-orange to a deep orange during the swaying (Figures 15, 16).

The Stretch Display of the male differs from that of the Common Heron in several respects: (1) the display of *cinerea* is performed in two separate motions, an initial one when the bird stretches its head and neck upward, then a second when the bird lowers its bill, head, and neck down and back, and bends at the heel joints, so that the whole body is lowered toward the nest; the same display is made by the Green Heron in one smooth movement without bending the heel joints; (2) during part one of the display of *cinerea,* the bird emits a soft "hu" or "huh" call, then utters a soft, gurgling "oooo" call during part two, whereas the single "aaroo" song of the Green Heron is given during the swaying movements; (3) the scapular plumes of *cinerea* remain depressed (see plate in Holstein, 1927, reproduced in Verwey) while they are erected to a full fan shape in the Green Heron (see Figure 16). The modified jugular plumes of *cinerea,* judging from the plates, are partly erected during the display, whereas these feathers are not modified and are not erected during the Green Heron's display; (4) eye-bulging and iris color changes, rare in the Green Heron, are not recorded for *cinerea.*

Both sexes of the Green and Common Herons show the Stretch, but the form, time of appearance, function, and orientation of the display differ between the sexes of the Green Heron. The female Green Heron assumes the final Stretch Display position exactly as does the male, that is, in one smooth movement, but the form of the female's display differs in the following respects: (1) the female, to my knowledge, never erects her scapular plumes; (2) the female never emits any sound at all; (3) the sways of the female are reduced in number and amplitude, or are completely absent; and (4) I have

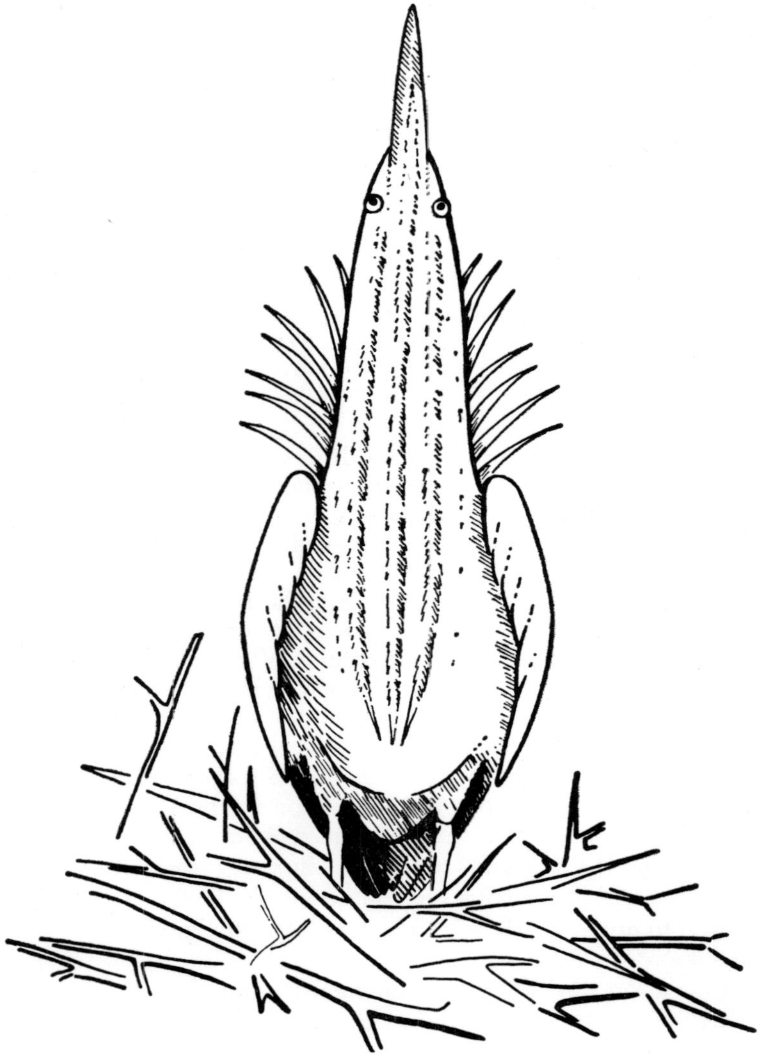

FIGURE 16. The Stretch Display of the male Green Heron as seen from the front.

FIGURE 17. The Stretch Display of the female Green Heron.

never observed eye-bulging or iris color changes in her display (see Figure 17).

The female shows the Stretch Display only *after* the male has permitted her to enter the nest, while the male gives his Stretch Displays, with rare exceptions, *before* allowing the female to enter the nest for the first time. The rare exceptions occur when the Stretch Displays appear during nest relief and just before copulation. The Common Heron differs strikingly, showing Stretch Displays frequently during nest relief.

The Stretch Display has three functions in both sexes, two of which are shared equally by the male and female, that is, nest relief and synchronizing nest construction. In the female, the display is also a pre-copulatory invitation or solicitation posture, typically but not invariably shown prior to copulation. I have seen a male show the Stretch Display immediately before copulation on only a few occasions (Table 13). The male's Stretch Display also invites the female to enter the nest; that is, the male signals to the female that hostilities are subdued or over.

The orientation of the display differs in the two sexes, the female showing her back to the male, while he characteristically turns his swaying breast and head toward the female. The scapular plumes of the male form a fan behind his head and neck (Figure 16) and emphasize his display; that is, they assist in exaggerating the display and drawing attention to the swaying, cooing male. The fan does not function at all in the invitation display of the female. Her scapular plumes are not erected during the performance, and her body is turned away from the male (Figure 17).

I agree with Moynihan (1955b) that the causation of appeasement displays can be quite complex. In the typical Stretch Display, I feel that the sexual tendency predominates, and a low intensity form of the Stretch Display supports my belief. During the early stages of pair formation when males are "skowing" and females are "skeowing," males will occasionally perform a low intensity Extended Stretch Display which has the following characteristics: (1) the displaying bird raises his head and neck as in a typical Stretch Display, but the neck is *not* fully retracted; the bill, head, and neck are *not vertical,* and a peculiar kink is evident in the partly retracted neck; the alignment of bill, head, and neck is forward of the vertical by about 30–40 degrees; (2) the male raises his scapular plumes only slightly or not at all; (3) he sways from side to side only a few times; and (4) he does *not* give the "aaroo" call at all.

This Extended Stretch Display (Figure 18) expresses predominantly hostile tendencies because: (1) males frequently terminate an Extended Stretch Display with one or more rapidly repeated "skows"; (2) appearance of a female on the territory releases threat displays

FIGURE 18. The Extended Stretch Display of the male Green Heron.

and/or pursuing or supplanting; and (3) the Extended Stretch Display is shown on or off the territory. In the typical Stretch Display the male never "skows" at the end of the display, the male shows Crooked Neck Flight, Flap Flight, or Snap Displays when the female approaches the territory, and the male confines the majority of his Stretch perpormances to the area around the nest site. Briefly, typical Stretch Displays alternate with displays having fewer hostile components, while Extended Stretch Displays alternate with much more hostile behavior.

In the Stretch Display the chief heron weapon, the bill, is turned upward and away from the female, and the neck of the displaying bird is exposed. There is no crest-erection; in fact, the only feathers erected are the scapular plumes. Threat display in the Green Heron exhibits extensive plumage-erection, wing-waving, low, horizontal body positions, and is accompanied by harsh, rasping calls. In the Stretch Display, on the other hand, the feathers (with the exception of the scapulars) are depressed, the bill pointed up and away from the female, and the call is typically low and pleasant-sounding ("aaroo"). This call is found in no other Green Heron display, and it is confined to the male.

A male in the Stretch Display, with his lustrous plumage, dark lores, brilliant coral-orange legs, gentle sways, and gurgling calls, offers a signal quite the opposite to that of the bristling, rasping fighter of the Full Forward Display. The female must have a clear signal that she will be permitted to enter the nest after she has been repeatedly driven from it.

Another striking peculiarity of the Stretch Display is the individual variability shown by birds on Rulers Bar Hassock. This variability involves the number of sways, the number of "aaroo" songs, and the quality of the song. In 1955, I recorded ten or more Stretch Displays by each of thirty-four individual males. Of the thirty-four, thirty-one had soft songs, while three had harsh songs. No males shifted from one quality to the other during the courtship period; that is, a male remained soft or harsh. Table 4 lists the records of Stretch Displays for seven of the thirty-four males observed in 1955. Note that the number of sways and the number of "aaroo" songs varied from one to six. No individual, however, showed the entire range of variation (male 1–31 showed the extreme but concentrated on two sways with two songs per display). A complete movement from left to right was counted one sway, and the return movement from right to left was counted another sway. The "aaroo" song in the vast majority of displays is uttered during one movement, but male 1–12 sang once for every *two* sways. One male (1–15) in forty-one Stretch Displays invariably showed only two sways and two songs per display. Male 1–4 was exceptional in that he showed six sways and six songs per

TABLE 4
INDIVIDUAL VARIATION IN COMPONENTS OF THE STRETCH DISPLAY OF MALE GREEN HERONS.

Male	Song Quality	No. Stretch Displays	No. of Sways						No. of Songs					
			1	2	3	4	5	6	1	2	3	4	5	6
1-1	soft	66	1	45	20	–	–	–	1	45	20	–	–	–
1-10	harsh	44	–	40	4	–	–	–	–	40	4	–	–	–
1-15	soft	41	–	41	–	–	–	–	–	41	–	–	–	–
1-31	soft	38	–	34	2	1	1	–	–	34	2	1	1	–
1-13	harsh	21	–	12	9	–	–	–	–	12	9	–	–	–
1-12	soft	17	–	–	–	17	–	–	–	17	–	–	–	–
1-4	soft	10	–	–	–	–	–	10	–	–	–	–	–	10

display, but I have records on this bird for only one day. Most of the counts point to two, and the vast majority of my single observations and records of series of less than ten displays are also of two sways and two songs per display. In the section on Late Pairing Behavior, activity records are reproduced to show that as the intensity of Stretch Displays rises, the majority of the males rarely add more than one sway and one song per display.

I have no data on geographical variation in the Stretch Display, and the only records I was able to find published (Wheelock, 1906; Simmons, 1925) are not detailed enough for comparison.

Daanje (*op. cit.*) discussed the origin of the Stretch Display and suggested that the second phase of this display in *A. cinerea* is to be considered the deep phase of the jump, an intention movement of flight. Ritualization in both *cinerea* and *virescens* has exaggerated the movements in different ways. In *cinerea* the ritualization has emphasized the second part of the display, lowering of the body. In *virescens*, exaggeration has affected the side to side swaying movements. Both species give a pleasant, gurgling call in marked contrast to their threat calls, an example of the evolution of an appeasing sound.

Billing and other pre-copulatory behavior. Prior to the first copulation, both of the pair are tense, and a sudden movement releases low intensity hostile behavior, for example, crest-raising and Forward Displays. The tenseness is obvious: the female has just entered the nest, the male has just permitted another Green Heron to enter the *sanctum sanctorum,* and both birds are very wary, leading to rapid shifts in behavior.

Billing, where both birds extend their head and neck and gently grasp the mate's bill momentarily, together with feather-nibbling and bill-snapping, are preludes to the first bodily contact. Such brief contacts appear to reduce the attack tendency, inhibiting withdrawal responses and increasing the sexual tendency. Mutual billing, bill-snapping, and feather-nibbling have been recorded for *A. cinerea* by

Verwey, Witherby *et al.*, (1949), and Lowe (1954), but I have been unable to find any references for the Green Heron. Similar behavior has been reported for the Roseate Spoonbill, *Ajaia ajaja* (Linnaeus) by Beetham (1910) and Allen (1942), and for the White Stork, *Ciconia ciconia* (Linnaeus) by Schüz (1942) and Haverschmidt (1949).

Copulation. Mutual feather-nibbling and related behavior are prolonged prior to the first copulation, but after the first copulation precopulatory behavior is reduced to a minimum. During the nest-building or nest-repairing phase, when copulations are most frequent, the pair pass twigs and perform occasional mutual Stretch Displays for an extended period. The female suddenly gives a single Stretch Display (Figure 17) with her back turned toward the male. The male becomes greatly excited and moves toward the female, which is usually standing on the floor of the nest. At the male's approach, the female assumes a squatting position low in the nest, with her wings drooped out and down to the sides. The male enters the nest, nibbles the flank and/or neck feathers of the female for a few moments, then makes a slight jump-flight onto the female's back. Both sexes, especially the male, flap their wings to maintain balance.

Occasionally, the pair stands quietly side by side in the nest, then the female suddenly shows a rapid Stretch Display, squats low, awaiting the male, which then nibbles her head, neck, and flank feathers for a few moments prior to mounting. The vast majority of copulations take place in the nest, but occasionally a pair will copulate while standing on a branch near the nest. I have never seen Green Herons copulate away from the nest tree.

The copulation lasts from five to twelve seconds, ten seconds being most common. Typically, the male leaves the nest after copulating, perches nearby and preens for a minute or two, then returns to the nest to engage again in mutual feather-nibbling and billing. While the male is preening immediately after copulation, the female typically works on the nest, adding a twig, adjusting others, and so on. Several hours or longer normally intervene between copulations. I have not seen one copulation follow another immediately.

The male appears to be much more tense during pre- and post-copulatory behavior than is the female. In fact, the male at all times shows hostile behavior more easily immediately preceding and following close contacts between the sexes. The sight of the female engaged in nest-building or other non-hostile activities, however, apparently reduces his hostile tendencies and allows him to return to the nest and engage in mutual nibbling.

After the female has laid one or more eggs of her clutch, the pair typically copulates when the female has relieved the male on the nest. I saw no copulations after the last egg in the clutch had been laid.

I saw pseudo-male behavior by a female in copulation only once.

On this occasion the male had been adding twigs to the new nest for several minutes, and he had retired to a nearby branch to preen and shake out his feathers. The female flew into the tree suddenly, and the male adopted at once the Withdrawn Crouch. The female alighted on his back, wing-waving to maintain balance. She immediately flew to the nest, gave a single Stretch Display, squatted low in the nest with wings drooped, and waited. The male entered the nest and copulated with the female for eleven seconds.

Incomplete copulatory attempts have been described under the Stiff-necked Upright Display. I found no evidence for promiscuity, such as Meanley (1955) found in the Little Blue Heron, *Florida caerulea* (Linnaeus).

Twig-passing. Before the male allows the female to enter the nest, he does all the gathering, carrying, and weaving of twigs into the nest. The nest site is always selected by the male. Once the male has allowed the female to enter the nest, however, he gathers and carries most of the twigs and the female does most of the building. Low intensity Stretch Displays are common early in nest-building before, during, or after the presentation of a twig. Either sex may display first, but it is usually the female. These Stretch Displays usually include two sways, but neither sex utters any sound at all. Here is a typical sequence: the female inserts a twig in the floor of the nest as the male looks on from a nearby perch; the female performs a single, silent Stretch Display; a few moments later, the male walks and jumps to a neighboring tree, breaks off one twig, brings it back and passes it to the female standing in the nest. While the female is "tremble-shoving" (Lorenz, 1955) the twig into place, the male performs a single, silent Stretch Display on the rim of the nest. Later, both may show single, silent Stretch Displays side by side on the floor of the nest. This display and twig-carrying are repeated throughout the day for several more days, but after an egg or two has been laid, the bird incubating the egg(s) performs most of the nest-building activities. Stretch Displays become infrequent and are confined typically to nest relief. Early nest-building activities in *B. virescens* are strikingly similar to those of *A. cinerea,* as described by Lowe (*op. cit.*).

When Green Herons take a twig, they do so by extending their head and neck fully out and down (in most cases), seizing the twig crosswise in the bill, then breaking it off by tugging and shaking motions. Twigs are carried back to the nest in any position. These twig-grasping movements are the basis for the ritualized Snap Display.

EARLY PAIRING BEHAVIOR

In this section I shall describe the behavior and appearance after arrival on the breeding areas until the first sexual displays are shown. There is no clear cut division of pairing behavior into hostile and sexual phases. In general, however, there is a decided decrease in hostility as pairing progresses. Hence, I use the term "pair formation" broadly to embrace all of the behavior of the two sexes *after* the female has ceased to roam and confines her attention to a single male, until the first copulation (*cf* Baerends and Baerends, 1950).

As far as I was able to determine on Rulers Bar Hassock, Green Herons are unmated when they arrive on their breeding grounds. However, the time between the first arrival and the laying of the first egg in the three years 1954, 1955, and 1957 was so short that I suspect more association of the two sexes than "migratory sociality" during the northward flight.

Year	Arrival Date of First Green Heron	Date of First Egg
1954	April 20	May 2
1955	April 19*	April 30
1957	April 20	April 29

*Two Green Herons were noted on April 16, 1955, but they flew north out of sight. These records refer to birds that remained on Rulers Bar Hassock.

Flying Around and Related Activities. Flying Around behavior has been described as the first reproductive activity after arrival (p. 38). I am convinced that the females seek out the males, even at this early stage. They fly past and land near a perched male repeatedly, even though repelled again and again by threat displays and/or Pursuit Flights and Supplanting Attacks. The females are attracted to the males, not to the territories.

Figure 19 records the movements (diagrammatically) of one bird that I took to be a male. The observations were made on April 21, 1957, the day following the first arrival of Green Herons on Rulers Bar Hassock. I saw the lone male first leaving 2 BA (that is, breeding area No. 2; see map, Figure 1) at 1342 hours. It flew unhurriedly to a perch in the northern part of 1 BA, where it preened briefly. Flying Around, perching, looking around, and so forth, were repeated until 1435, when the bird flew from the central part of 1 BA to the neighboring feeding area. It walked slowly north for a few minutes, then returned to the northern part of 1 BA. The bird did not make a strike while it was on the feeding area, and it is my impression that the bird was uninterested in feeding. I made many similar observations of this early Flying Around behavior in 1954, 1955, and 1957.

Figure 20 shows the first observed contact between a male and a female earlier in the day on April 21, 1957. I have used a recording technique similar to those of Hinde (1953) and Bastock and Manning (1955). The notes read from left to right and from the top to the bottom, as one would read a type-written page. Table 5 is a key to all of the symbols used in Figure 20 and similar figures. Note that the female "skeowed" once in flight (1051) just before landing near

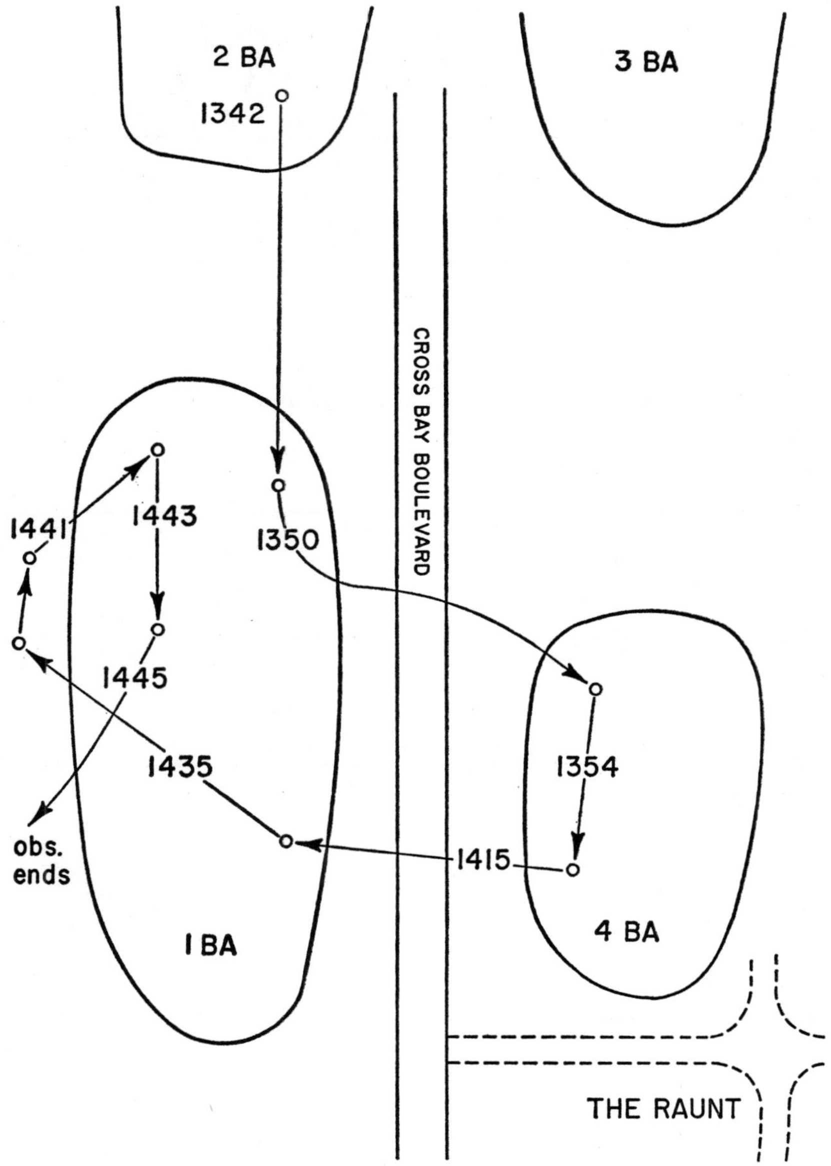

FIGURE 19. Flying Around behavior of a lone male Green Heron prior to settling on a territory. All times are EDST.

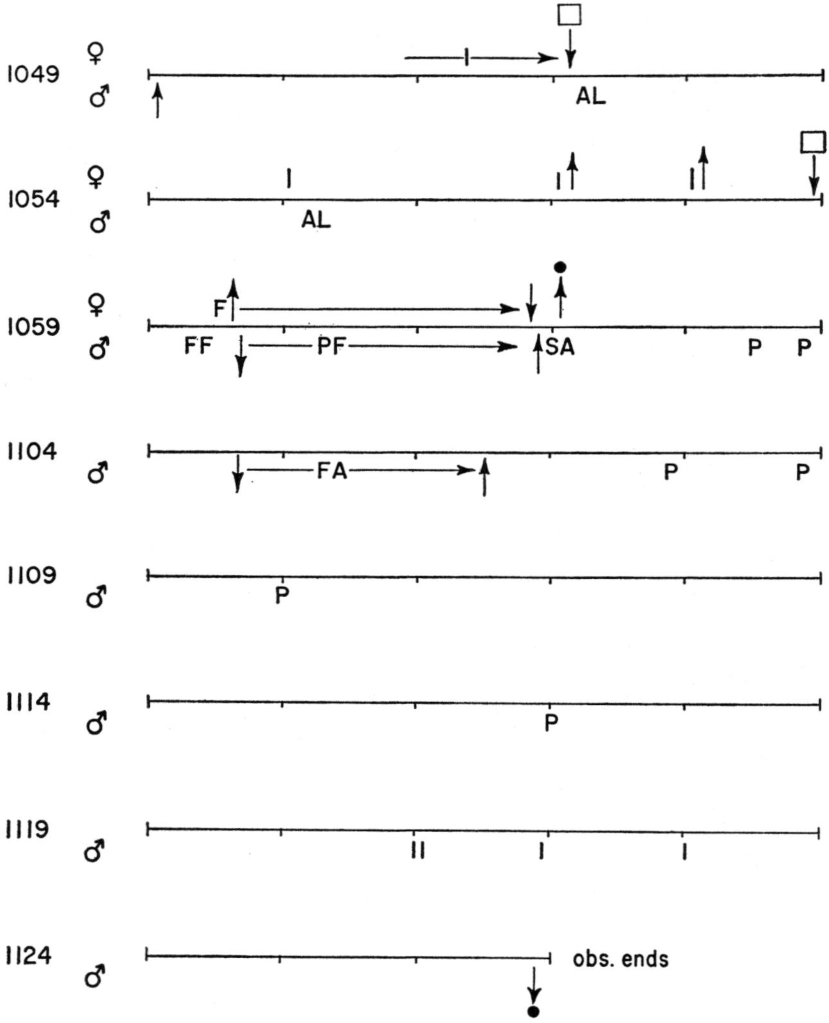

FIGURE 20. An "ethogram" showing the behavior of the male and female Green Heron on the first contact at the beginning of the breeding season. A key to the symbols used in this and all following "ethograms" is to be found in Table 5.

TABLE 5

A KEY TO SYMBOLS USED IN THE "ETHOGRAMS" OR BEHAVIOR RECORDS.
ALL TIMES ARE EDST.

1012 |——1——2——3——4——5 minutes

Symbol	Meaning	Symbol	Meaning
1012	time of observation	⊙	nest
♂	male	□	nearby tree
♀	female	1♂	male no. 1, etc.
♂ I	one skow	1♀	female no. 1, etc.
♀ I	one skeow	——PF——►	pursuit flight
R	stretch display	——I——►	skow or skeow in flight
S	snap display	————►	bird flying by
AL	alert posture	——MH——►	Marsh hawk flies by
F	forward display	I I	infrequent "skowing" or "skeowing"
FF	full forward display	I II II I I	increased frequency
FL	flap flight display	IIIII IIIII	a rapid burst
P	preening	↓ / ↑	lands / takes off
BS	bittern stance	●↓	lands out of sight
FA	flying around	●↑	takes off, obs. terminated
O	nest tree		

the perched male. After "skeowing" a third time, the female flushed, landed closer to the male, and "skeowed" a fourth time, at which the male threatened her with a Full Forward Display. The female responded with the lower intensity Forward Display, flushed, and was pursued by the male, which supplanted her when she attempted to land (1102). After the female flew out of sight, the male flew around the area, perched, "skowed" four times, then flew away.

As more migrants arrive on the breeding areas, Flying Around, Pursuit Flights, threat displays, and advertising calls increase in frequency and intensity as a result of more inter-individual contacts. The initial behavior is purely hostile, even before the males have settled on territories and the females have started roaming.

Roosting. Both males and females roost on or close to the ground. Upon their arrival, they roost anywhere in the breeding area. Later, when the males have settled on territories, the birds roost only on their territory. I noted roosting sites every night, and during the early part of the breeding cycle no site was occupied on successive nights. Green Herons land high and fly or jump directly down to the roosting spot. They rarely walk down branches to the selected site. When leaving the roosting spot, they normally fly up, but they walk up branches to the take-off spot more frequently than they walk down branches to the roost. Seibert (1951) also states that Green Herons fly directly down to their roosting spots.

Hostile behavior during early roosting involves an occasional threat display, apparently to maintain "individual distance" (Hediger, 1942; Conder, 1949). Fights are rare and usually result from an attempt by a late arrival to roost in an occupied site. Such fights are brief, and it is usually the intruder that withdraws. Settling on roosts at this early stage is typically rapid and uneventful because there are few herons, and males have not yet established territories.

Once the males have established territories, hostile behavior becomes more pronounced and more prolonged; and threat displays, fights, and Supplanting Attacks become common. This activity produces much milling around, changing of roosts, and a prolongation of settling time.

On the day that I heard the first "skowing" in 1955 I also observed that many of the males which had advertised during the day also "skowed" from their roosting perches. I heard such calls even long after dark when most of the birds were settled for the night.

Territory. I shall follow Noble (1939) in defining territory as "any defended area," with the exception that I shall restrict my discussion to breeding territories. Feeding territory has been mentioned briefly in the section on hostile behavior.

After the brief Flying Around and/or Pursuit Flight stage, a marked change appears. Now the males, following these aerial maneuverings, orient their return flights more and more to a particular portion of the breeding area. The females now shift to roaming. That is, they now fly from one ill-defined male territory to another, not at all restricting their attentions to one particular male. The male's response is purely hostile.

At this stage advertising calls begin in earnest: "skow" (male) and "skeow" (female). At first, the boundaries of the male's territory are ill-defined, but the initial area is *always* larger than the subsequent territory defended by the pair after pair formation. The initial large size of the territory depends upon a number of factors: (1) not all males enter the advertising stage at the same time because migration is protracted and few males occupy the large breeding areas; these early arrivals can establish large territories because border clashes are relatively uncommon at this time; (2) males settling on a territory for the first time patrol this area, advertising from a number of song posts, and this additional elbow room is apparently advantageous in attracting a female. In other words, a male that makes himself conspicuous by moving about and advertising from many song posts stands a better chance of attracting a female than a male in more crowded quarters. It may be that the conspicuous calling and moving around attracts the roaming female; (3) a larger area will have more perches, thus allowing a female to stay on a male's territory longer than if she was forced to land near the nest site, where she would be vigorously attacked. Once the pair is formed the large territory has no further advantage. These points will be treated in greater detail later.

As the season progresses, the number of song posts used decreases sharply, partly due to the arrival of additional males, and partly as a result of the female's behavior (Figure 21). Several things should be noted in this figure: (1) the male accepted my observation blind readily, "skowing" directly over my head; (2) at least one old nest was located on the initial territory; (3) the male used the nest tree as a song post, in addition to other trees; and (4) as the territory was reduced in size, the male "skowed" from fewer and fewer song posts, but the nest tree was always one of these. Eventually, the male only "skowed" from the nest tree or the nest platform itself.

A variety of evidence indicates that the nest, if an old one is available, or the nest site, is the primary object around which territorial behavior revolves: (1) regardless of how much the territory is eventually restricted in size, the nest or nest site is never abandoned; (2) whenever a male patrols or leaves the territory, he almost invariably alights in the nest tree first upon returning; (3) the most intense of the male's threat displays are almost always performed from the nest platform; (4) the song post most frequently used, even when many posts are used, is the nest or nest tree; (5) almost without exception, nests that survive from the previous breeding season are used by the earliest males; and (6) a female, when visiting a male's territory during his absence, will show the greatest number of escape components when closest to the nest. When peering in toward the nest, or when perched in the crown of the nest tree, the females appear to be very tense and nervous.

Tables 6, 7, and 8 substantiate these points with quantitative data. In all three seasons (1954, 1955, 1957), as soon as a male was observed "skowing" from an old nest platform, this fact was recorded and the

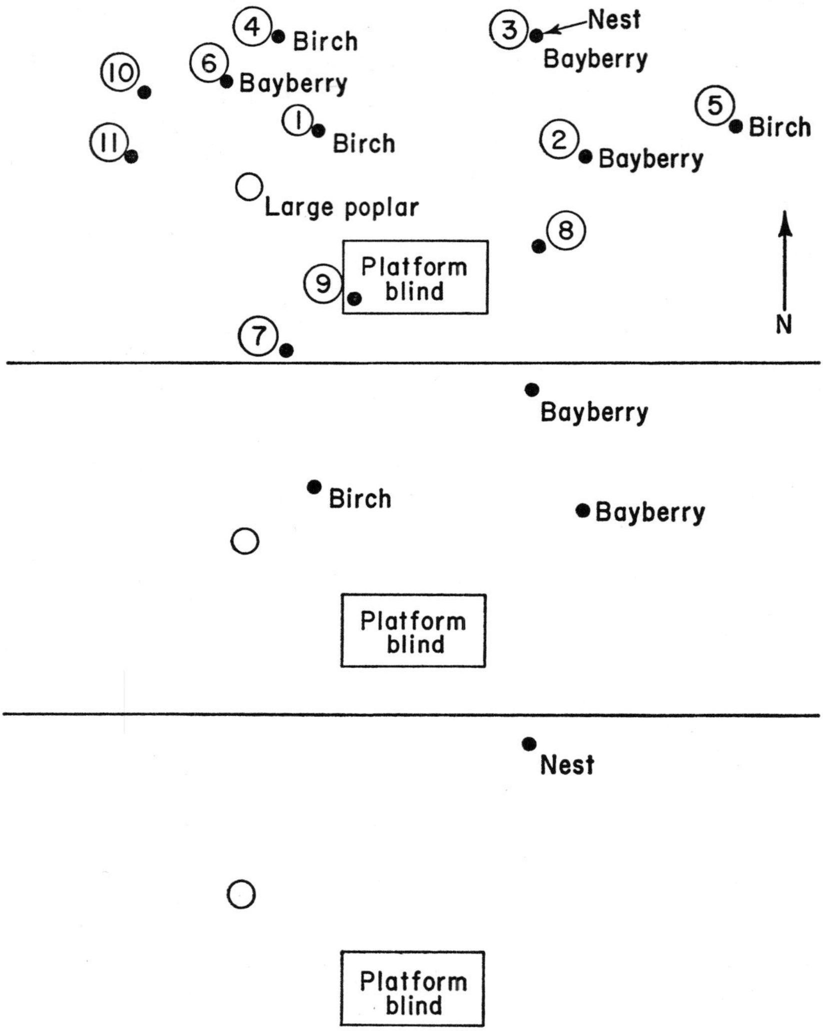

FIGURE 21. "Skowing" perch reduction in the male Green Heron. The upper third of the figure represents the initial "advertising" area where the male "skowed" from eleven song posts, one of which was a nest later used by the pair. The male's behavior at this stage was almost purely hostile. The middle third of the figure represents a later period; the male is now beginning sexual displays. Note the sharp reduction in number of sites used for "skowing." The lower third of the figure represents the final "skowing" phase where the male advertises from *one* site—the nest.

TABLE 6

SIGNIFICANCE OF THE NEST TREE IN THE TERRITORIAL BEHAVIOR OF THE MALE GREEN HERON, RULERS BAR HASSOCK, BREEDING SEASON OF 1955.

Male	No. Observed Return Flights to Territory	No. Return Flights to Nest Tree
1-1	201	159
1-2	147	129
1-10	56	48
1-17	83	63

TABLE 7

SIGNIFICANCE OF THE NEST TREE AS A SONG POST IN THE TERRITORIAL BEHAVIOR OF THE MALE GREEN HERON, RULERS BAR HASSOCK.

Date	Male No.	No. "skows" from Song Post						Nest Tree
		1	2	3	4	5	6	
April 23, 1955	1-1	1	2	1	3	–	–	47
" 23, 1955	1-2	3	–	–	–	–	–	41
" 24, 1957	1-17	3	3	4	5	8	17	196
" 27, 1957	1-23	2	1	–	–	–	–	95
May 14, 1957	3-3	10	13	7	–	–	–	286

TABLE 8

RE-USE OF NESTS AT THE BEGINNING OF EACH BREEDING SEASON, GREEN HERON.*

Year	No. Nests Surviving	No. Surviving Nests Re-used	No. Nests Tagged at End of Breeding Season
1954	unknown	unknown	63
1955	59	53	102
1956	unknown	unknown	none
1957	68	67	none

*All data from Rulers Bar Hassock. No observations were made in 1956; the nests reused at the start of the 1957 season were tagged survivors from the end of the 1955 season.

nest was tagged. In every case, eggs were laid in the surviving tagged nests. Because no birds were color-banded, I cannot rule out nest-abandoning or territory-shifting, but males closely observed every day of the breeding season gave no impression of shifting and abandoning territories.

Although behavior revolves around the nest, it is not clear why an old nest or a possible nest site should be so critical. There seems to be an abundance of trees and building materials for nests. The importance of the nest may be tied up with the fact that twig-passing plays a limited role in courtship. In contrast, twig-passing is much more important in the mutual displays of Black-crowned Night Herons during courtship. Black-crowns have a much more mobile courtship, and old nests are infrequently used for first nestings. Also, the male Green Heron does a considerable amount of repairing or building prior to pair formation, while both sexes build the bulk of the nest together in the case of the Black-crowned. Nest removal and transfer experiments are necessary to determine the significance of the nest in the Green Heron's territorial behavior.

As more males arrive and start to settle on territories, border clashes increase sharply, and the boundaries of the territories become defined temporarily. But the boundary of a Green Heron's territory shrinks steadily in size. Several factors are concerned with this shrinkage: arrival of new males, behavior of the females, the rough synchronization of neighboring males' behavior, and the shift from hostile to sexual displays. This shift in motivation is, I believe, the key factor.

Figure 22, though diagrammatic, represents the actual territorial behavior of seven male Green Herons during the breeding season on Rulers Bar Hassock in 1955. At stage 1, males A, B, C, and D have started territorial behavior at about the same time. Note that A's territory is large compared to his subsequent area. The behavior of the males at this stage is purely hostile: "skowing," threatening, patrolling, and fighting at the borders. Each male had at least one old nest on his territory; these nests are represented by a large black dot. At stage 2, males A, B, C, and D are now "skowing" much less frequently and from fewer song posts, and, more important, each of these males is engaged in activities, both hostile and semisexual, with one particular female. Note that males E and F have arrived, and settled on large initial territories, and are at the purely hostile stage. Contacts between E and F and the four earlier males, ABCD, are infrequent, but ABCD do threaten and attack all male intruders coming into their reduced territories. The factor which reduces E-ABCD or F-ABCD conflicts at this stage is that when males A, B, C, and D return to their nest trees after foraging or other flights, they fly *directly* to their small territories. This and their reduced "patrolling" minimizes the number of contacts between them and E and F. Too much fighting on the part of ABCD at this stage would seriously hamper their relationships with the females; hence, the advantage of the direct flights. Allen (1951) suggests that such direct return flights by *nesting* Snowy Egrets reduce clashes between neighbors, and his photograph shows adults with eggs and young. This type of direct return flight occurs even earlier in the breeding cycle of the Green Heron.

At stage 3 in Figure 22, males ABCD are at the nest-building or nest-repairing stage, while males E and F are now negotiating with individual females. Male G has arrived on the scene, and he starts to patrol a large initial territory. It is of interest and importance to note that even though G engaged in fewer actual fights than did the other males which arrived earlier, G's territory also shrank steadily in size.

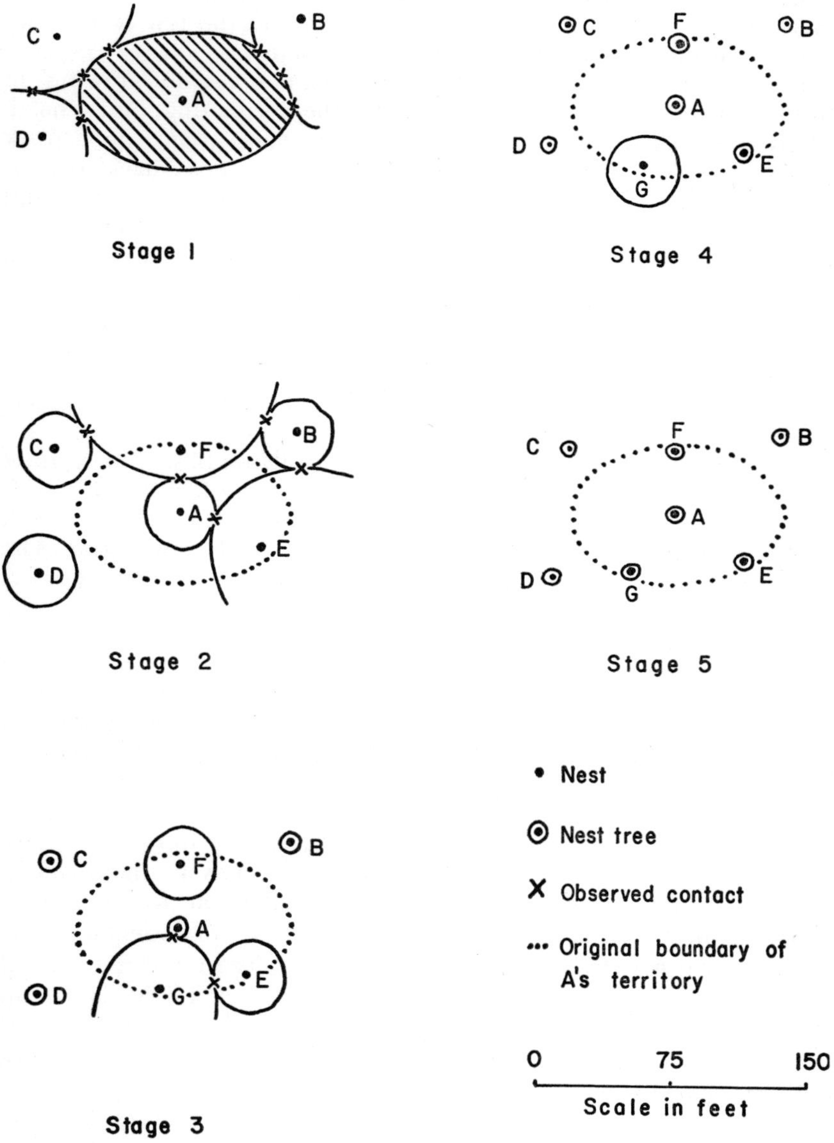

FIGURE 22. Territorial behavior in the Green Heron. Full discussion in text.

Stage 5 completes the picture. On the area initially defended by A (Stage 1), four pairs of herons are now breeding, and the respective pairs now defend *only* the nest, its contents, and the few feet immediately surrounding this one indispensable object.

I have seen incubating birds completely ignore a male "skowing" in the crown of the tree five feet above the incubating bird's head, behavior which would have resulted in immediate attack just a week previous. Again, these incubating birds seem to sneak back to their nests, running a gauntlet of threatening males. During flight, and shortly after alighting in their nest tree, these birds show many escape or general alarm responses, that is, rapid crest-erection and depression, sleeked plumage, and tense appearance, with no calls or threats. The males at the hostile stages seem to ignore these trespassers after the latter land in their own tree. It may be that failure of the incubating birds to respond to threat displays while in flight, and their direct, tense return flights are signals which inhibit attacks, i.e., the trespassers exhibit a type of "submissive behavior."

How the intrustion of late arrivals on well established territories takes place has presented an intriguing problem. I feel that the shift in behavior from hostile to sexual displays has been little explored, and that this shift is the key to the problem in the Green Heron. The rigid territorial boundaries which are maintained long after pair formation, reported in so many species, especially Passerines, are simply not found in the Green Heron.

Functions of territory. The recent *Ibis* symposium on territory in birds (Hinde *et al.*, 1956b) clearly emphasized the diverse functions of territory. With the Green Heron, territory has several functions, and these vary from stage to stage in the breeding cycle. These are intracolony functions, not long-range or ultimate, as suggested by Lack for *A. cinerea* (1954a, b). First, the aggressiveness and advertising of the male, together with patrolling, that is, his mobility in the fairly large initial territory, make him more conspicuous to the roaming females. Second, once a female confines her attentions to a particular male, the pair then has an area which facilitates pair formation. The early mutual activities of the pair include a large number of aerial displays (for example, Circle Flight, Crooked Neck Flight, and Flap Flight), and too sharp a reduction in territory size at this stage would make contacts with neighbors too frequent, thus seriously interfering with the behavior of the pair. Later, the two sexes confine their displays to a small area around the nest or nest site, and the displays of the male, especially, are stationary ones, such as the Stretch and Snap. Third, most copulations take place on the nest or only a few yards from it, and apparently the vigorous defense of the nest site prevents interference with copulation. Fourth, the territorial behavior of the male secures for the pair a nest site or a previous year's nest.

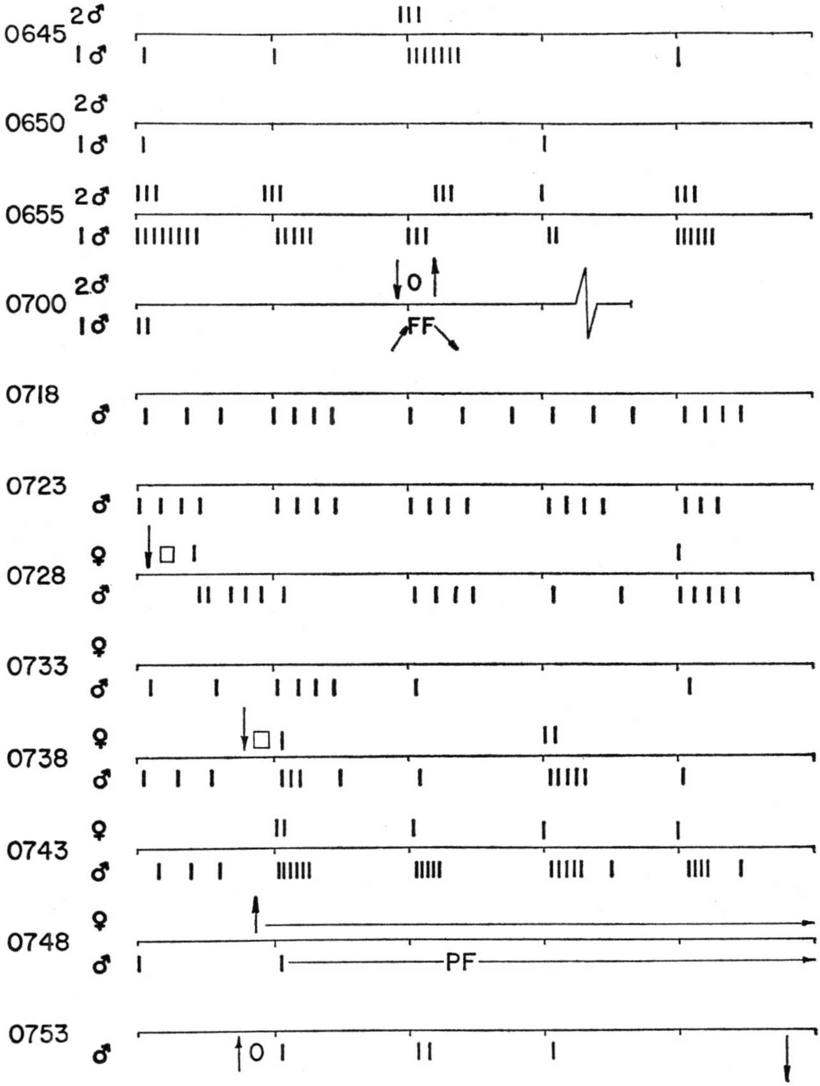

FIGURE 23. The early territorial advertising of the Green Heron. Note the complete absence of sexual displays. See Table 5 for a Key to the symbols.

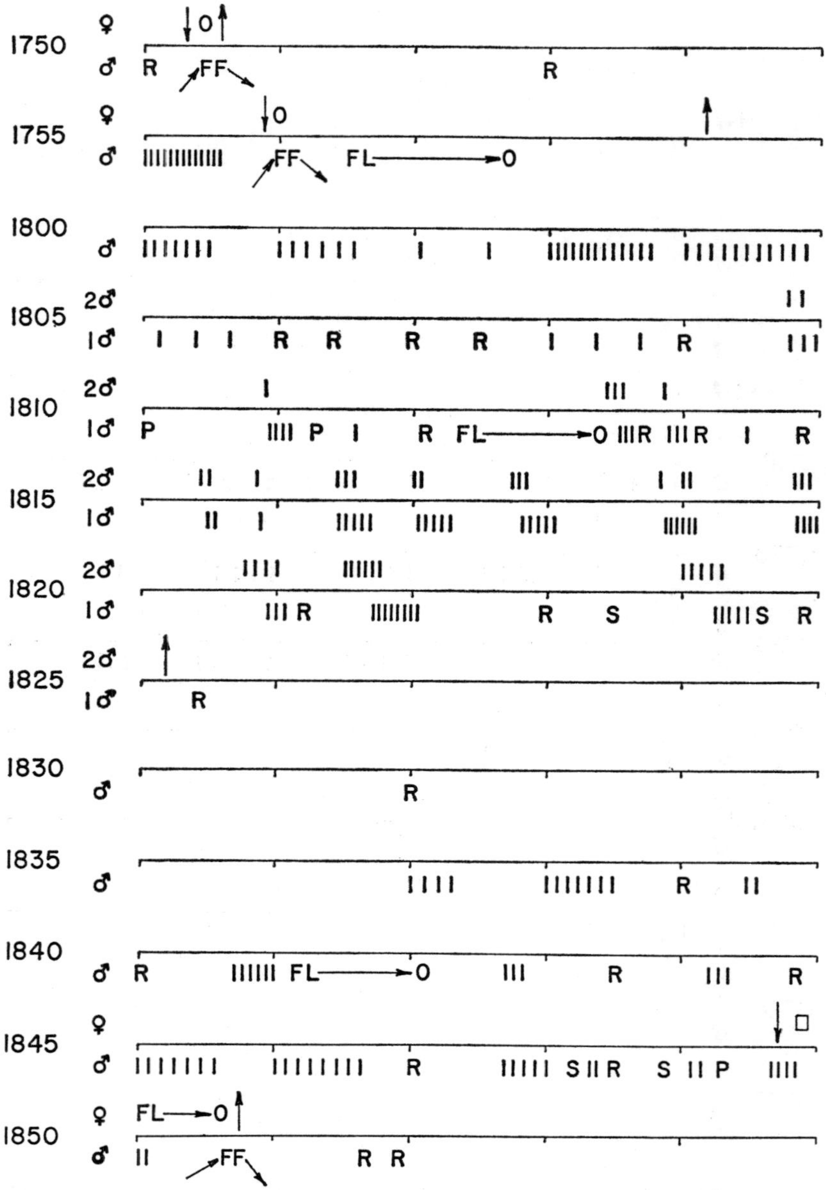

Figure 24. An "ethogram" showing the effect of "skowing" on neighboring territorial male Green Herons. Note particularly the "skow-trading" between 1809 and 1824 hours. The diagonal arrows at 1750, 1756, and 1850 indicate that the male walked up to the crown of the nest tree, threatened the female with a Full Forward Display, then walked back down to the nest after the female flew away, as shown at 1750, or remained in the crown of the nest tree (1756), or flew past the nest tree in a Flap Flight Display.

Mutual hostile behavior. Once the males have settled on territories, both males and females become highly vocal. Periods of "skowing" are interrupted by short preening bouts, twig-grasping, and an occasional Pursuit Flight when a roaming female approaches too closely. This early advertising of the males is shown in Figure 23. This male was "skowing" from the floor of an old nest, and he remained there until about 0800, when a female landed in the crown of the nest tree. The female left immediately after a Full Forward Display by the male, which then pursued the female. Note that the male remained in the nest even when the female landed at the borders of his territory at 0739; also that "skeowing" by the female stimulated the male to "skow" (e.g., at 0739, 0746, 0747, and so on).

Shortly after a Pursuit Flight the male typically returns to his territory and resumes "skowing," as shown in Figure 23. The frequency of "skowing" varies considerably, from as low as one "skow" per hour to sixteen "skows" per minute. Towsend (1928) records a frequency as high as ten "skows" in a quarter of a minute, but thirteen "skows" in half a minute was the only record of such a burst I was able to make. Bursts of 10–14 "skows" per minute were quite common, however. "Skowing" frequencies vary with the time of the day, "skowing" of neighboring males, and presence or absence of a female. "Skowing" males definitely stimulate each other, and one such bout of "skow trading" is shown in Figure 24. Such "skow trading" territorial males soon learn the exact location of their neighbors and their song posts. A "long distance threat" function, however, is another possibility, because the males on distant territories also engage in such "duels." The threat function of the "skow" is indicated by the reaction of a male that landed in the territory of a rival when the intruder was unaware of the owner's presence because he was concealed by reeds. A single "skow" from the owner released immediate withdrawal on the part of the intruder, even though the owner was still out of sight (cf, Lack, 1953, on the threat function of song in the English Robin, *Erithacus rubecula* (Linnaeus)).

The advertising call of the male clearly attracts females, and I believe that female roaming can be explained partly on the basis of this attraction. From a high wooden platform I could watch a female fly from one "skowing" male to another, and it was quite evident that the female settled, at least temporarily, on the fringes of the territory of that male which engaged in a "skow-skeow" bout with her.

The female's advertising call seems to be a stimulus that releases the male's "skow," thereby informing the female of the exact number, location, and "behavioral disposition" of all males. Figure 25 shows that the female's "skeow" is stimulating to the males. Note the bursts of "skowing" after each "skeow" from the female. Also note that the frequency of the male "skowing" increased sharply after the arrival

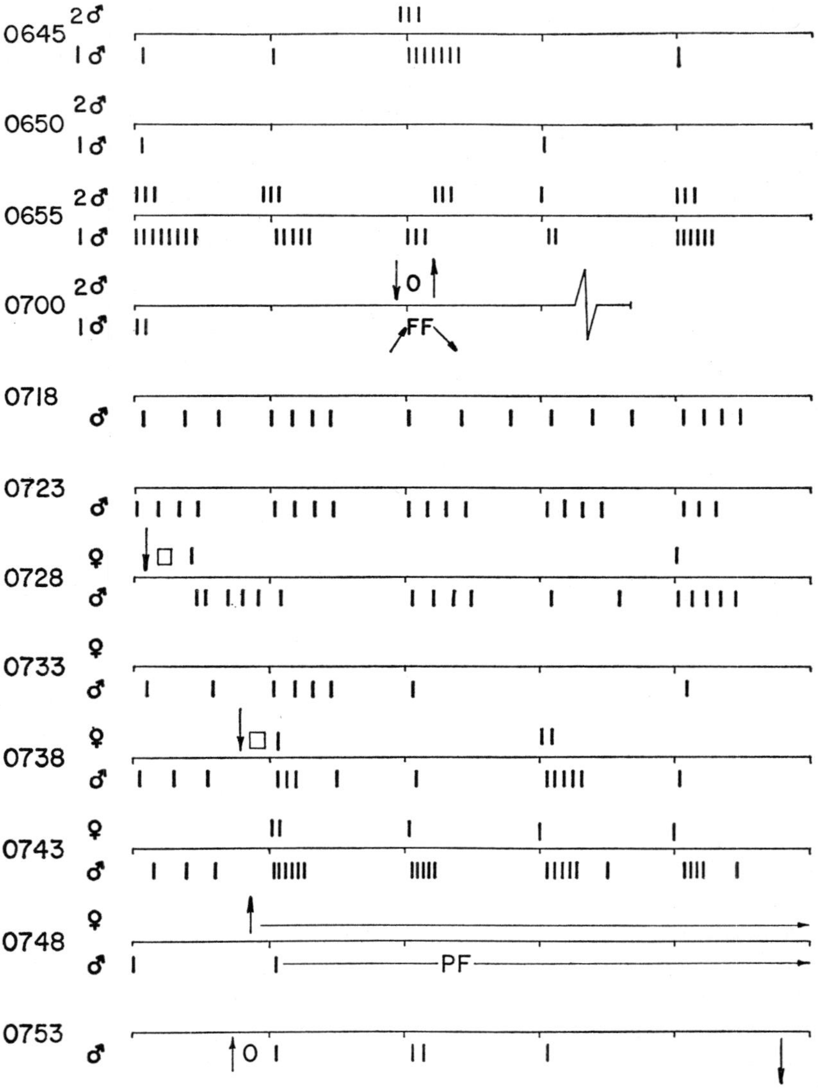

FIGURE 25. An "ethogram" showing the effect of female "skeowing" on the "skowing" frequency of a territorial male Green Heron. Note particularly the period between 0718 and 0748 hours.

of the "skeowing" female. When a female is threatened or attacked and then leaves the territory of the hostile male, she "skeows" from the fringes of his territory or from a neighboring, momentarily unoccupied territory, as if to test his readiness for her return. If the male is ready to respond he "skows," thus releasing a return flight by the female. If the male does not "skow," however, the female remains quietly perched for a few moments, then repeats her "skeow." Finally, if the male still fails to "skow," the female may return to the close vicinity of the male anyway, or she may fly away. Apparently the females approach the males right from the start, and their persistence in returning again and again, even though repeatedly attacked, is a striking feature of their behavior. Experiments might establish what I suspect, that the "skeow" call modifies the reaction of the male even though the change is not observed at once. Briefly, the conspicuous differences in the calls are hardly accidental.

At the same time that the male shows the first sexual displays, typically low intensity Stretch and Snap Displays, the females stop roaming and confine their attentions to these particular males. Figure 26 shows the behavior of a "pair" at this stage. Note that the female, perched on the territory of the male, but not close to the nest tree where the male was displaying, attempted to land in the nest tree at 1630, but that she was repulsed by the threatening male, which then pursued her. The male returned to the nest tree at 1632 and resumed "skowing." The female again attempted to land in the tree occupied by the male, was forced to withdraw, and returned again at 1742, shortly after a Stretch Display by the male. The female's "skeowing" released two short bursts of "skows," then a series of "skows" and Stretch Displays by the male. At 1811, the female attempted to land once more in the nest tree and was threatened by the male; but note that the male showed a Flap Flight Display this time rather than a Pursuit Flight (1812). Figure 27 shows the later behavior of the same pair. Note the sharp increase in frequency of Stretch and Flap Flight Displays; also that the male's behavior was still hostile, especially when the female tried to land in the nest tree (1150), or when she flew too close to the nest tree (1148, 1204, and 1229).

It is evident from my discussion that the "skow" is not inhibited by the arrival of a female. Verwey states that the *Liebeslockruf* or "mating call" of the Common Heron is instantly inhibited by the arrival of a female. By contrast, the arrival of the female Green Heron definitely stimulates the male to "skow" at a much greater rate than previously. Only after the male begins sexual displays does "skowing"' decrease. Another striking difference between *cinerea* and *virescens* is that the mating call of *cinerea* is restricted to males, while both sexes of *virescens* have advertising calls. It is possible that the advertising call of the female Common Heron, if such a call exists, has been overlooked by

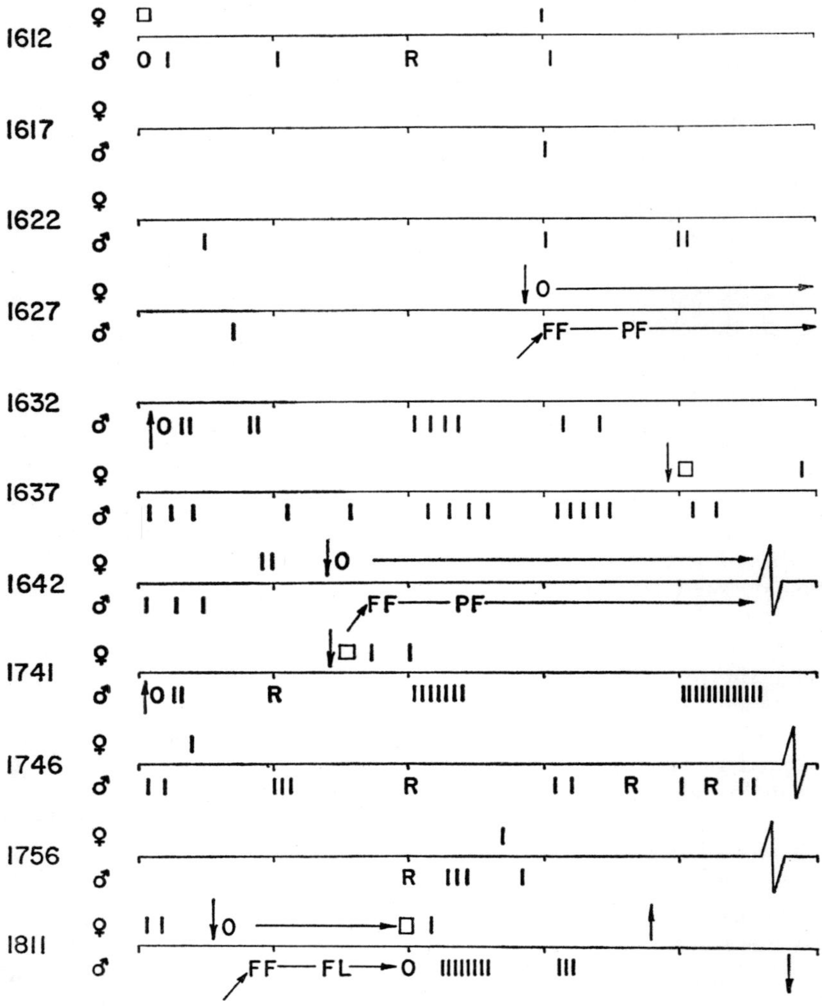

FIGURE 26. An "ethogram" showing the first appearance of male sexual displays and cessation of roaming by the female. Note the persistent return behavior of the female, even though she is repeatedly repulsed by the male. The diagonal arrows at 1630, 1643, and 1812 indicate that the male walked up to the crown of the nest-tree and displayed a Full Forward Display.

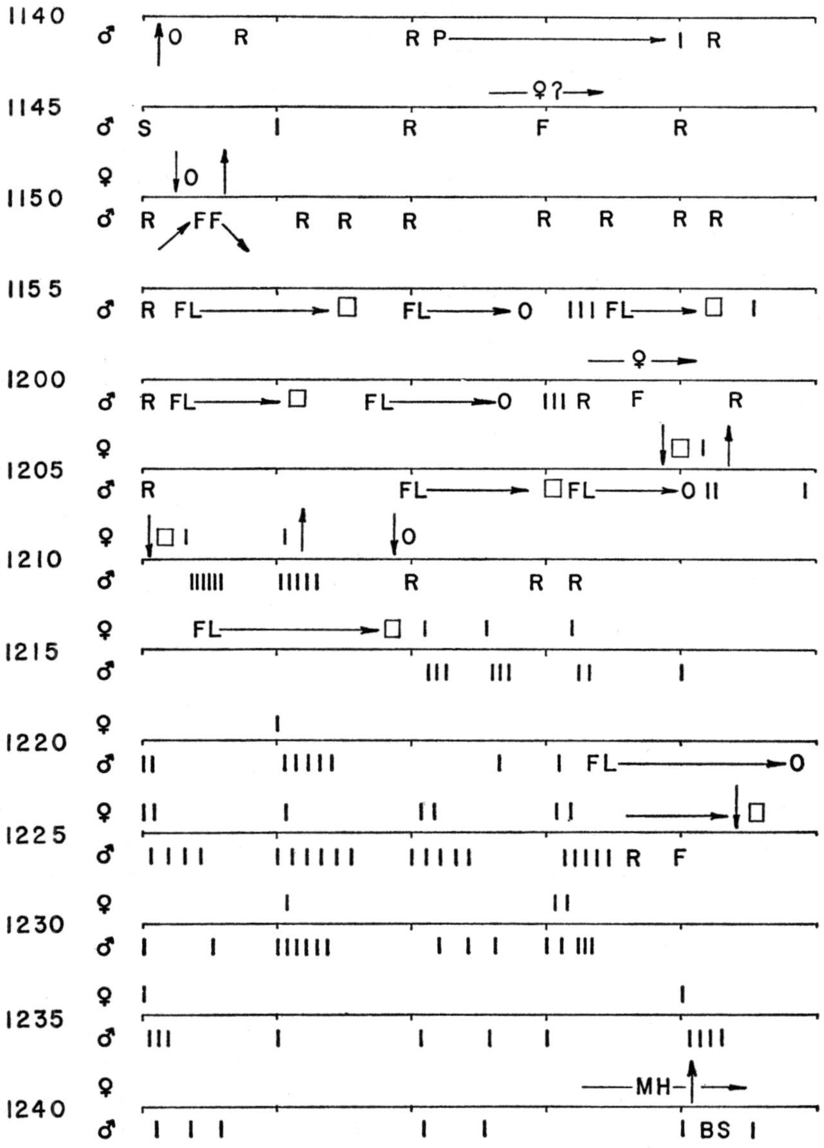

FIGURE 27. An "ethogram" showing the sharp increase in sexual displays of the male Green Heron after the female has ceased to roam. The diagonal arrows at 1150 indicate that the male walked to the crown of the nest tree, showed a Full Forward Display, then walked back down to the nest when the female flew away.

European observers because many of these displays are fleeting in nature.

The previous emphasis on male hostility during the early pairing behavior does not mean that females are showing primarily sexual behavior at this time. Frequently, when the females attempt to land in the nest tree occupied by the male, the females show just as intense threat displays as the males. Occasionally, the sexes engage in vicious fights in the crown of the nest tree before the female leaves, but the female *always* retreats. In general, the behavior of the female is less aggressive than that of the male because an important function of her behavior is to stimulate the male to react in a sexual rather than in a hostile manner.

Nest-building before pair formation. As mentioned earlier, the female takes no part in nest-building or nest-repair until *after* pair formation. The male makes extensive repairs to an old nest or constructs the foundation of a new one. At first, the male may merely adjust the twigs in the old nest. Later, the male occasionally grasps, then quickly releases, twigs near the nest tree or in the nest tree itself. Then, a twig may be broken off, only to be dropped. These dropped twigs may be retrieved repeatedly, only to be dropped after "mouthing." Finally, a twig is secured, brought to the old nest, and inserted with the

TABLE 9

MORPHOLOGICAL AND BEHAVIORAL CHARACTERS USED IN SEX DISCRIMINATION BY MALE AND FEMALE GREEN HERONS.

Character	Male	Female
Advertising call	"skow"	"skeow"
Site tenacity	tenacious	roams
Snap Display	present	absent
Stretch Display—time of occurrence	common before and during pair-formation	only shown after pair-formation
Stretch Display—vocal quality	with "aaroo" song	always silent
Twig-grasping	common prior to pair-formation	absent prior to pair-formation
Nest-building	starts repair on old nest or lays foundation for new nest prior to pair-formation	no nest-building activities until after pair-formation
Stiff-necked Upright Display	present	absent
Soft-part coloration	brilliant	duller than male
Plumage "luster"	lustrous	duller than male
Response to threats of territory-holding male	threatens, fights, or withdraws	if interested in male, withdraws and returns repeatedly

TABLE 10

EFFECT OF SOME ENVIRONMENTAL FACTORS ON THE BEHAVIOR OF NEWLY ARRIVED GREEN HERONS. RULERS BAR HASSOCK, EARLY SPRING, 1955.

Date 1955	Temperature °F.				Wind Velocity (mph)			Behavior
	Max.	Min.	Av.	Diff.*	Av.	Max.	Diff.*	
Apr. 20	64	45	55	—	7.6	22	—	First Flying Around, First Pursuit Flights
" 21	61	43	52	–3	6.8	16	–0.8	Increased Flying Around, Pursuit Flights
" 22	69	48	59	+7	7.5	27	+0.7	First "skowing," First fighting
" 23	69	49	59	0	4.3	15	–3.2	Great activity, including first Stretch Display, SA's, Flap Flight Displays
" 24	57	47	52	–7	7.5	17	+3.2	Occasional Flying Around
" 25	48	44	46	–6	12.5	29	+5.0	An occasional "skow"
" 26	45	42	44	–2	15.4	22	+2.9	No activity
" 27	54	44	49	+5	11.8	23	–3.6	An occasional fight
" 28	56	46	51	+2	15.1	22	+3.3	Activity increases, nest-repair noted
" 29	65	48	57	+6	16.6	33	+1.5	Greatest activity to date
" 30	73	51	62	+5	9.2	20	–7.4	Increased activity, first egg laid
May 1	65	49	57	–5	6.7	17	–2.5	Very active, additional nests and eggs

*Average difference from the previous day.

characteristic heron "tremble-shoving" movements. Frequently the male will sit in the nest up to one hour, even during the early phases of courtship. While sitting in the nest, the male adjusts the twigs of the floor and rim. He continues until the nest is almost completely repaired.

Sex discrimination. The first response of a territory-holding male Green Heron to intrusion by either sex is always hostile, ranging from stationary threat display to pursuit and fighting. Pair formation is eventually effected, however, so the two sexes must some how tell each other apart. I believe that Green Herons discriminate sex by the same characters that I used, that is, a combination of morphological and behavioral characters (Table 9). The most reliable are sexual displays, position in copulation, and role in nest-building.

Influence of weather. Tables 10 and 11 show that in both 1955 and 1957 there was an initial burst of activity, then a lull, and then a renewal. In both years the period of lessened activity occurred shortly after a sharp drop in the average daily temperature, or during a brief period of low temperatures. The decrease in activity was most noticeable in 1955 when the average daily temperature dropped steadily

TABLE 11

EFFECT OF SOME ENVIRONMENTAL FACTORS ON THE BEHAVIOR OF NEWLY ARRIVED GREEN HERONS. RULERS BAR HASSOCK, EARLY SPRING, 1957.

Date 1957	Temperature °F.				Wind Velocity (mph)			Behavior
	Max.	Min.	Av.	Diff.*	Av.	Max.	Diff.*	
Apr. 20	65	49	57	—	10.3	26	—	First arrival on area
" 21	82	61	72	+15	19.2	34	+8.9	First Flying Around, first Pursuit Flights, first "skow-skeowing"
" 22	65	49	57	–15	13.7	26	–5.5	Occasional Flying Around, Pursuit Flights, first low intensity Flap Flight Display
" 23	68	49	59	+2	7.1	17	–6.6	Increased Flying Around, Pursuit Flights
" 24	80	58	69	+10	9.0	16	+1.9	Greatly increased "skow-skeowing," threat displays, fighting, occasional Flap Flight
" 25	59	50	55	–14	10.1	20	+1.1	Greatly decreased activity
" 26	78	50	64	+9	9.1	18	–1.0	Slight increase in activity
" 27	72	54	63	–1	7.6	17	–1.5	First Stretch Display, increased activity, prolonged nest-repair
" 28	77	53	65	+2	6.6	16	–1.0	General increase in activity
" 29	75	61	68	+3	13.0	26	+6.4	First egg laid
" 30	71	57	64	–4	20.1	36	+7.1	Much activity, nest repair, more eggs laid
May 1	80	58	69	+5	20.6	43	+0.5	Much activity, additional nests with eggs

*Average difference from the previous day.

for three days. Although the average wind velocity increased during this period, it is apparent that the drop in temperature was the crucial factor because, following the rise in temperature, activity was renewed even though the wind continued to rise. The same increase in activity with rise in temperature was noted in 1957, and then, too, the average wind velocity increased during the active period. When temperature is rather constant, however, the wind is an important factor. During periods of blustery weather, the earliest arrivals remain inactive and quiet, perched low in the protected parts of the breeding areas. In the later stages in the breeding cycle the wind reduces mutual advertising displays.

During rain storms most of the birds stop flight displays and go to protected perches on the breeding areas. On several occasions, however, I saw threat displays during relatively heavy downpours. In general, low temperatures, high winds, and moderate to heavy rains reduce activity.

Coloration of soft-parts. "Soft-parts" refers to the unfeathered portions of the body, i.e., the bill, lores, iris, legs, feet, and skin surrounding the eye. An early reference to the "red" legs of *virescens* is in Wilson and Bonaparte (1878). Christy (1929; cited in Roberts, 1936), after flushing a Green Heron in the month of May, 1928(?), states, "He was extraordinarily handsome, or so he seemed to me, and particularly I remarked his coral-orange legs." Allen (1934) and Pough (1951) both note the orange-red color of the legs and feet of the male during the breeding season, and Allen's book contains a color plate by George M. Sutton (opp. p. 141), which shows well the brilliant coloration of the legs and feet. However, Allen's statement (1934:144) that "hers remain yellow or greenish-yellow" does not hold for the females I observed in Rulers Bar Hassock, eastern Massachusetts, Florida Bay, and Puerto Arista, Chiapas, Mexico. Roberts (1936) and Pough (*op. cit.*) similarly restrict the orange-red coloration to the males. Leg-color changes during the breeding season have been noted in the closely related *Butorides striatus* in Surinam by Haverschmidt (1953), and by Stanford (1936) and Smythies (1953) in Burma.

Table 12 lists the breeding and non-breeding soft-part coloration of both sexes, with brief remarks on plumage. The only data I have

TABLE 12

SOFT-PART COLORATION OF MALE AND FEMALE GREEN HERONS.

Area	Male	Female
	Non-breeding	
bill	brownish-black, basal part of lower mandible yellowish	same
lores and orbital skin	greenish-yellow	same
legs and feet	bright yellow to yellowish-green	same
iris	yellow	same
general plumage characteristics	greenish (may appear bluish) wings and back, maroon neck, greenish-black crest	same
	Breeding	
bill	lustrous blackish-brown to shiny jet-black	same, but less intense
lores and orbital skin	intense blue-black but variable	dull blue-black, highly variable
legs and feet	brilliant coral-orange	same, not as intense
iris	yellow to orange, deep orange on occasion	yellow
general plumage characteristics	rich, lustrous green wings and back, rich maroon neck, glossy greenish-black crown; modified scapular plumes have silvery tone	similar coloration, but much duller

relative to geographical variation in soft-part coloration is the observation that the male Green Herons of the Puerto Arista area of Chiapas, Mexico, are the most brilliantly colored I have ever seen. Individual variation was marked, especially on Rulers Bar Hassock.

I have found no references to color changes in the bill, lores, and iris for *virescens* or *striatus*. In *virescens* these parts change in color only during a brief period during courtship. Iris color changes noted in some of the displays are especially transitory. The coloration of the lores, too, changes rapidly, especially in the female, where it also shows much individual variability.

These soft-part color changes enhance and emphasize the various hostile and sexual displays. The colors of both sexes fade rapidly to the winter color after egg-laying. The female's color has faded completely by the time the last egg has been laid. At this time the male's coloration is dull, but not as "washed-out" as that of the female. By the time the young hatched I could not recognize the sex by any differences in coloration.

LATE PAIRING BEHAVIOR

Once the female begins to confine her attentions to one male, her roaming ceases, and the male's behavior rapidly becomes more and more sexual. Instead of only making "skows" along with threats and Pursuit Flights, the male begins to perform more and more Stretch, Snap and aerial displays, such as the Crooked Neck and Flap Flight Displays. Figure 28 shows the behavior of the "pair" at this stage. Note the increase of sexual displays and the sharp reduction in "skowing," except when a neighboring male "skows." Note also the sharp reduction in female "skeowing" which is no longer functional. Most of the female's behavior consists of attempts to land in the nest tree, and short flights from tree to tree around the displaying male. The female now lands in the nest tree occupied by the male, and she remains perched there even though threatened with high intensity hostile displays (1743, 1756, and 1917). The female is not always tolerated in the nest tree, however, as can be seen at 1750 and 1850 in the same figure.

Part of the function of the female's short flights, both normal flights and display flights (e.g., Circle Flights), seems to be to get the male to follow her in such a way that the female can land first in the nest tree upon returning. The female is frequently successful in this ruse, but her stay in the nest tree is usually brief, especially when she tries to perch too close to the nest. As the mutual flights increase in frequency, however, the female is permitted to stay in the nest tree for longer and longer periods.

A common sight at this time is that of a male performing one Stretch or Snap Display after the other, from the floor of the nest, while the

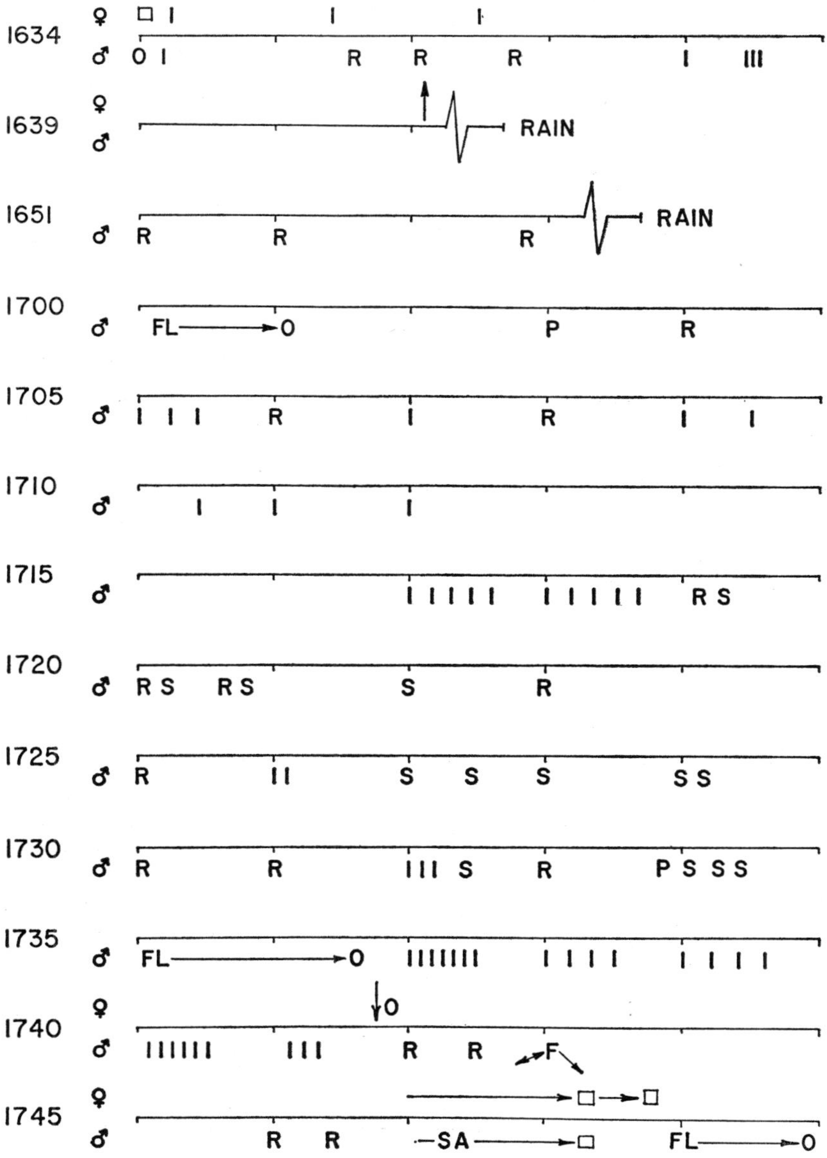

FIGURE 28. An "ethogram" showing the late stages of pair formation behavior in the Green Heron. NOTE: Figure 28 consists of three continuing "ethograms." The double-headed arrow at 1743 indicates that the male started to walk up to the crown of the nest-tree (where the female was perched), hesitated, showed a Forward Display briefly, then walked back down to the nest.

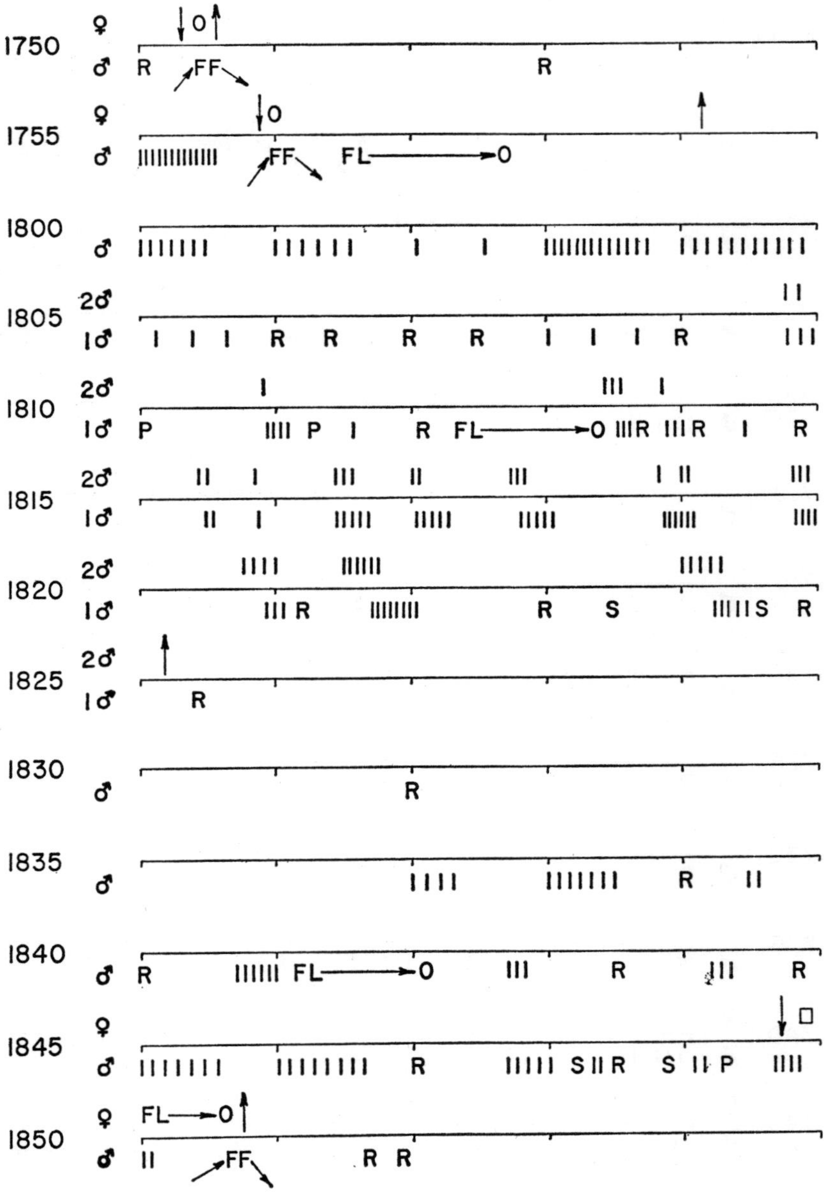

FIGURE 28. (Continued).

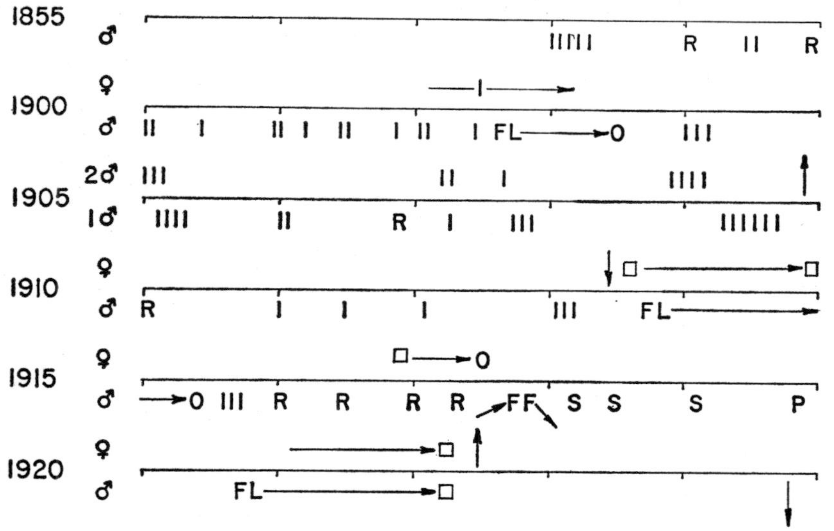

FIGURE 28. (Continued).

female, perched in the crown of the nest tree, peers down intently at the swaying, softly "aarooing" male. Sometimes the female tries to approach the male during such a display, only to be repulsed, but she is rarely forced to leave the tree. The male simply stops the sexual display, threatens her briefly, and when the female has retired to the crown of the tree, he resumes his sexual displays. When the male is away, the female shows an active interest in the unoccupied nest, peering at it intently, walking slowly down a branch toward the nest, then retreating. Her anxiety is clearly indicated by rapid tail-flicking and crest-raising, and by her reluctance to enter the nest.

Finally the crucial moment comes when the female is permitted to enter the nest for the first time. If any act can be considered as the turning point in the pair formation of the Green Heron, it is this. I have witnessed this performance on many occasions, and the behavior of the two sexes is almost invariable. Both are obviously tense, but the male seems to be more highly agitated because his tail-flicking and crest-raising are more pronounced. If the male is standing in the nest before the female enters it, he shows a low intensity Forward Display as she approaches. The female, which is moving cautiously toward the nest, shows a fleeting Forward in return, but her feather-erections are extremely slight. As the female enters the nest, any sudden movement by either bird instantly triggers rapid feather-bristling by both sexes. Typically, the male *leaves* the nest as the female steps in, and he perches on a nearby branch, peering intently at the female in the nest. The male characteristically moves about the surrounding branches in a low, crouched position, tail-flicking and crest-raising steadily. As soon as

TABLE 13

FREQUENCY OF VARIOUS PRE-COPULATORY ACTIVITIES BY ONE PAIR OF GREEN HERONS (PAIR 1–13), RULERS BAR HASSOCK, SPRING OF 1955.

Date	Period of Obs.	Bill-snapping			Billing		
		Mutual	Male	Female	Mutual	Male	Female
May 9	8 hrs.	–	–	–	2	–	1
" 10	3 "	1	–	–	1	–	–
" 11	3 "	2	–	–	1	–	–
" 12	11 "	6	–	–	6	–	–
*" 13	12 "	2	–	–	5	–	–
" 14	4 "	–	–	–	–	–	–
*" 15	10 "	–	–	–	–	–	–
" 16	2 "	–	–	–	–	–	–
*" 17	8 "	–	–	–	–	–	–
*" 18	1 "	–	–	–	–	–	–
" 19	3 "	–	–	–	–	–	–
Total 11 days	65 hrs.	11	–	–	15	–	1

Date	Period of Obs.	Feather-nibbling			Stretch Displays			Copulations
		Mutual	Male	Female	Mutual	Male	Female	
May 9	8 hrs.	1	–	–	1	–	1	1
" 10	3 "	1	1	–	1	–	1	1
" 11	3 "	2	2	–	–	–	1	1
" 12	11 "	7	–	1	1	1	1	–
*" 13	12 "	2	20	3	–	–	3	1
" 14	4 "	–	–	–	–	–	–	–
*" 15	10 "	–	6	1	–	–	–	1
" 16	2 "	–	–	–	–	–	–	–
*" 17	8 "	2**	–	–	–	–	–	–
*" 18	1 "	1**	–	–	–	–	–	–
" 19	3 "	–	–	–	–	–	–	–
Total 11 days	65 hrs.	16	29	5	3	1	7	5

*Deposition of one egg.
**Nest relief ceremony.

she has gained admittance to the nest, the female starts to fidget with the nest twigs, but she does so nervously, picking at a twig here, then looking at the male, turning around in the nest, and then touching another twig.

If the male is perched in the crown of the nest tree at the female's first entrance into the nest, he adopts the Forward Display and stalks toward the female. Occasionally, the male gives a few very soft "raaah-raaah" calls as he nervously approaches the female, and as above, any sudden movement releases brief hostile displays. This is a trying period for both sexes, and the nervous moving about the nest and nearby branches, fidgeting with nest twigs, bristling, and watching each other, continue for the rest of the day. If one or both of the pair leaves the territory (that is, to feed or if flushed by a sudden noise), the sequence is repeated upon their return. Regardless of the intensity of the male's threat displays, the female now remains standing in the nest, and for a brief period, the observer gets the distinct impression that now it is the male that is the intruder!

After the first entrance of the female into the nest all male "skowing" and Snap Displays, all female "skeowing," and all aerial displays cease completely. After a few encounters as described, the first pre-copulatory activities appear, such as bill-snapping, billing and feather-nibbling. Pre-copulatory behavior is prolonged before the first copulation, and then these mutual activities decrease in frequency and intensity with each copulation, apparently because of "familiarization."

Table 13 shows the frequency of pre-copulatory activities for one pair from the first entrance of the female into the nest until the completion of the clutch (four eggs). In general, bill-snapping and billing, which have hostile components, drop off sharply at the time the first egg is laid. These are mutual displays, while feather-nibbling becomes more of a male display and Stretch Displays more of a female display. Both of these latter displays, however, are shown mutually. Most Green Herons carry over feather-nibbling into nest relief ceremonies. Also, the pair occasionally show brief, always silent Stretch Displays at nest reliefs, which can hardly be called "ceremonies" because they are so brief, and because so little ritualization is involved. This contrasts sharply with the case of species such as *A. cinerea* and *A. herodias*, where elaborate nest relief ceremonies occur after egg-laying. As incubation progresses, Green Herons eventually relieve each other without any display whatever. The relieving bird enters the nest quickly and settles on the eggs as the relieved bird rises and leaves.

Following nest-building and egg-laying sexual behavior disappears suddenly and completely, and the pair enters the parental phase of the reproductive cycle. Even during the period of egg-laying, the pair spend little time in purely sexual activities, as shown in Table 14. Several activities shown are of interest. Note that both members of this pair (1–13) spent a great deal of time away from the nest before egg-laying, which is rather atypical for the Green Heron. Other mated pairs on Rulers Bar Hassock spent many hours in the nest together, standing quietly side by side. This companionship lasts until the deposition of the second egg, when the incubation routine of "one on, one off" starts. Note in Table 14 that the attentive periods on the first and second eggs are intermittent, that is, there are fairly long periods during which the eggs are not covered by either bird. With the laying of the third egg, however, a regular incubation rhythm is established in which the male incubates mostly in the midday and midnight hours, while the female sits during the early morning and late evening hours. This schedule, however, was highly variable from pair to pair on Rulers Bar Hassock.

TABLE 14

BEHAVIOR OF A NEWLY-FORMED PAIR OF GREEN HERONS (PAIR 1–13), FROM FIRST ENTRANCE OF FEMALE INTO NEST UNTIL COMPLETION OF CLUTCH, RULERS BAR HASSOCK, SPRING OF 1955. ALL TIMES ARE IN MINUTES.

Date	Period of Obs.	Pre-copulatory and Related Behavior	Preening		Nest-building			Incubation		Time Away From Nest	
			♂	♀	♂♀	♂	♀	♂	♀	♂	♀
May 9	480	30	20	5	17	5	6	—	—	390	420
" 10	180	12	5	7	5	30	—	—	—	120	150
" 11	180	26	4	—	—	5	—	—	—	120	150
" 12	660	41	34	60	20	19	3	—	—	540	540
*" 13	720	63	3	20	14	20	28	150	90	450	510
" 14	240	—	10	—	—	12	—	180	—	30	240
*" 15	600	29	9	60	2	34	2	360	60	165	450
" 16	120	—	5	—	—	3	—	105	—	7	120
*" 17	480	**2	17	4	3	7	4	420	53	30	420
*" 18	60	**1	—	2	—	2	1	45	9	13	48
" 19	180	—	6	—	—	—	—	165	—	5	180

*Deposition of one egg.
**Nest relief ceremonies.

Egg-laying dates and time of deposition are shown for four pairs in Table 15. To the best of my knowledge, no eggs were laid by any Green Heron on any of the New York City study areas outside the period 0500 to 1100. Note that the first egg is laid later in the morning than the remainder of the clutch; this is very characteristic. It was also characteristic for the female to be relieved by the male shortly after deposition of each egg. In every nest at which I observed the deposition of the first egg, the male settled on the egg without hesitation as soon as the female left.

The egg-laying intervals shown for pair 1–13 in Table 15 are characteristic for the birds breeding on Rulers Bar Hassock in all three years of this study. The first and second eggs were *always* laid two days apart in those nests checked in detail, but the other intervals varied with different pairs. In the majority of nests, the third egg was

TABLE 15

EGG-LAYING INTERVALS AND TIME OF DEPOSITION FOR FOUR NESTS, RULERS BAR HASSOCK, SPRING OF 1955. ALL TIMES ARE EDST.

Nest Number	Date of Laying Egg Number					Time of Laying Egg Number				
	1	2	3	4	5	1	2	3	4	5
1–12	May 11	May 13	May 14	May 15	May 16	1014	*	0621	*	0610
1–13	May 13	May 15	May 17	May 18	———	0938	0700	0630	0650	——
1–40	May 29	May 31	June 1	June 2	———	0716	0652	0605	0629	——
1–41	May 27	May 29	May 31	June 2	———	0830	0805	0746	0751	——

*Exact time of laying not observed, but both eggs thus marked were laid prior to 1200 hrs.

laid two days after the second, and all other eggs followed at one-day intervals. Pair 1–41 in Table 15 were most unusual in that all eggs were laid at two-day intervals.

Before pair formation Green Herons pay practically no attention to other species in and around the territory, but a marked change occurs shortly after the pair has been formed and *before* any eggs are laid. Then and for the remainder of the nesting phase both show intense hostile behavior toward all Passerines, but especially toward the two most important local egg predators, the Grackle and the Fish Crow. The threats and attacks also alert all incubating and brooding birds within earshot.

Part II

COMPARATIVE BEHAVIOR OF OTHER NORTH AMERICAN HERONS

Introduction

An enormous number of publications on animal behavior, especially bird behavior, have appeared in the past two decades. This writing stems primarily from the stimulation provided by the numerous theories and hypotheses of Konrad Lorenz and Niko Tinbergen. One of the many results of this "school of comparative ethology" is the renewed interest in, and the application of ethological findings to, systematics.

One of the major findings of the comparative ethologists is that evolutionary relationships are reflected in behavior patterns as clearly as in morphology. Furthermore, in many groups where morphological criteria alone have given two or three alternate classifications, behavioral characters have led to unequivocal decisions. As a result the taxonomist who formerly relied solely on structural characters now has at his command in many groups a body of behavioral information to assist him in solving taxonomic problems. The radical revision of the family Anatidae by Lorenz (1941) and Delacour and Mayr (1945, 1946) used behavioral information extensively.

The first task of the ethologist is to make an inventory of all the behavior patterns of each species being studied. This descriptive phase is essential before any thought can be given to experimental analyses of isolated problems. Much effort has already been wasted by experimentalists who considered this purely descriptive phase trivial. Aside from the compilation of behavior catalogues, comparative ethological studies provide information on the causation, function, and evolution of behavior patterns. Answers to these three problems for any given display may be completely obscure when such a display is observed in but one species.

Recent work in "systematic ethology" has centered on the lower taxa, that is, species and genera. The behavioral characters so far described, analyzed, and used in assessing relationships of species and genera are the so-called "Fixed Action Patterns" shown during threat and courtship activities. Great emphasis is placed on hostile and sexual displays. These displays are excellent subjects for study because they are frequently very striking and highly characteristic of a given species. Furthermore, when groups of species are closely related, the ethologist usually has little difficulty in determining those displays which are homologous.

Ethological studies require a great deal of time to be thorough and comparative, as do most evolutionary studies. Most ethologists, moreover, find that the intensive study of one species is necessary before an extensive study of a group can be planned. I shall use the Green Heron as a basis for comparison throughout the remainder of this paper. In Part II of this study the species are treated following the plan adopted for the Green Heron, but none of them have been studied in as great detail as the Green Heron.

The behavioral data gathered to date only permit provisional suggestions as to systematic relationships. Morphological studies, such as Bock's (1956), in which the student has museum skins readily available for most of the species, have been much more broadly comparative than my own research. Final decisions must await more detailed studies of the behavior of the majority of non-North American species of herons.

GREAT BLUE AND GREAT WHITE HERONS

Recent contributions to the systematic status of the Great White Heron (*Ardea occidentalis* Audubon of the A.O.U. Check-list of North American Birds, 1957) confirm Mayr's view that it is conspecific with the Great Blue Heron, *Ardea herodias* Linnaeus (Mayr, 1956; Meyerriecks, 1957b; Meyerriecks and Meyerriecks, 1958). No evidence in the literature, or field observations, suggest significant behavioral differences; hence, I treat them as color phases. Where appropriate, I refer to the Great White Heron as "Florida Bay *herodias*" and to the Great Blue Heron as "mainland *herodias*."

METHODS

All observations of Florida Bay *herodias* were made in the Florida Keys, especially Cotton Key (see map in Meyerriecks, 1957b). Most of the observations of mainland *herodias* were also made in the Florida Bay area, but important data were also gathered on Rulers Bar Hassock. I include important observations of both color phases made in the Florida Bay area by Robert P. Allen, and where appropriate I refer to the important study of Michigan *herodias* by Cottrille and Cottrille (1958).

The same techniques used for the Green Heron studies were employed, but many important observations were also made from an open boat and from an exposed spot on U.S. Route 1, the main road through the Florida Keys.

STUDY AREAS

Observations of mainland *herodias* on Rulers Bar Hassock were confined to migration and feeding behavior because breeding of the Great Blue Heron on Long Island, New York, is unknown (Cruickshank, 1942).

Florida Bay is roughly triangular in shape, and it is bounded on the north by the Florida mainland, on the south and east by those keys traversed by U.S. Route 1, and on the west by the Gulf of Mexico. I did most of my field work in the shallow eastern portions of the bay (rarely more than ten feet deep). This area is dotted with numerous keys (see map in Meyerriecks, 1957b). Broad reefs, fully exposed or only partially covered with water, run from key to key, and hundreds of individuals of seven species of herons can easily be compared on these feeding areas.

DISTRIBUTION AND DESCRIPTION

The white birds have a very restricted range, being known only from Florida Bay, Cuba, Isle of Pines, Jamaica, and the coast of Yucatan. The blue birds, however, range as breeding birds throughout

North America, from southern Alaska, Alberta, and New Brunswick in the north, to southern Mexico, the West Indies, and the Galapagos Islands in the south. The closely related Cocoi Heron, *Ardea cocoi* Linnaeus, replaces *herodias* throughout South America, while the Common Heron, *Ardea cinerea,* replaces it throughout Europe, Asia, parts of Africa, and Madagascar. On the smaller keys both phases nest solitarily or in small groups (2-3 pairs) in the red mangroves, the only suitable vegetation. The larger keys support many more breeding pairs, which characteristically nest in the black mangroves that make up the interior of the island, rarely in the red mangroves on the shore. The mainland birds may nest singly or in very large colonies.

Ardea herodias is the largest North American heron. It is approximately fifty inches (127 cm.) long with a wing spread of about six feet (183 cm.).

The white birds are immaculate and have short occipital plumes (which may be absent); and short scapular and jugular plumes. The blue birds are bluish-gray on the back and wings; the primaries are darker than the rest of the wings; forehead and crown are white, with the feathers on the side of the head black; the blue birds have long, black occipital plumes; scapular and jugular plumes are longer than in the white phase; the brownish-white throat is streaked with black; thighs are chestnut. The sexes of both phases are indistinguishable in the field except during the courtship period. Mayr (1956) has discussed the significance of bill and plume lengths in the two phases.

MAINTENANCE ACTIVITIES

Locomotion. No detailed studies of take-off, flight, or landing were made, but these differ only in minor details from those of the Green Heron. *A. herodias* does not normally erect its crest when taking off or landing as does the Green Heron. The large size of *herodias* presumably limits its agility.

The large body and long neck make the bird appear to fly more slowly than the smaller Green Heron, but published flight speeds are roughly the same (Wetmore, 1916, and Wood, 1933 for *A. herodias*; and Lowe, 1954, for *A. cinerea*). The flight of *herodias* is steady, strong, and direct, with deep wing beats, in contrast to the apparent undulating flight of *virescens.* I have seen the spectacular spiral descent flight of *herodias* in many places. These flights were similar to those of the Green Heron.

On short flights, or when alarmed near the nest, *herodias* characteristically flies with head and neck almost fully extended and the legs dangling below. Lowe describes similar flight behavior in *cinerea.* On long flights the neck is withdrawn in the characteristic heron S-shape, and the legs are held in line with the general body axis,

FIGURE 29. Great White Heron "peering over" while feeding.

trailing behind. On migration *herodias* flies higher than does *virescens*; the same was true of *herodias* when flying to and from distant islands in Florida. Unlike the Green Heron, *herodias* frequently glides during a sustained flight. Long glides before landing are common in both species, but those of *herodias* are quite spectacular at times.

Unlike the Green Heron, which sneaks along in a crouched position, *herodias* strides forward with body erect, head and neck extended. Each step is seemingly measured, the advancing leg being brought forward smoothly and unhurriedly. Toes are relaxed and then spread wide as the foot is carefully placed down. The bearing is majestic. Of the North American herons which I know, only *herodias* and the Common Egret, *Casmerodius albus* (Linnaeus), have these attributes, although they are approached by the Little Blue Heron.

Feeding behavior. The basic type of heron feeding behavior Stand and Wait, together with a slight modification Wade or Walk Slowly, is common to all herons. A feeding Great Blue Heron shows the basic method to perfection. Whether standing on some exposed shoal or among the arched prop roots of the Red Mangrove, these herons are the essence of immobility. They may hold the body upright, with head and neck fully extended at about a 45 degree angle, or they may crouch low, body held horizontally, the bill almost touching the surface of the water. The only movement is that of the large, yellow eyes as they follow every motion of the intended prey. The statuesque immobility is maintained for long periods, but eventually the bird either strikes or moves slowly away. From either position, crouched or erect, the strike is swift and direct.

Both phases of *herodias* show, as do all herons, a slight modification when they wade or walk slowly. Wading is slow and each step is measured; the body may be erect or low and horizontal. The latter position is typical just before a strike when wading forward. Peering over (Figure 29) is very common, as if the bird were trying to locate prey which has moved behind something. This attitude may be held for several minutes at a time.

In one modification of Wade or Walk the bird wades forward very slowly in an erect position and then suddenly extends and withdraws its wings about a foot in a short, rapid flick. Such wing-flicks may be repeated as many as five or more times, but usually two or three flicks in rapid succession are made, then the bird resumes wading. I have seen wing-flicking only on bright days, in open, shallow water, when each wing-flick created an obvious sudden shadow on the surface of the water. The function of wing-flicking is to startle prey. It is true that the shadow of the bird's body is always present on the surface, but this moves slowly, while the wing-flicks suddenly appear and disappear as two extensions of this shadow. I have seen wing-flicking in exactly the same fashion in the Common Egret and the Little Blue Heron. Wing-flicking is, I believe, a low intensity form of Open Wing Feeding, beautifully shown by the Snowy and Reddish Egrets and the Louisiana Heron, *Hydranassa tricolor* (Müller).

I have not seen hovering over the water, skipping over the surface, or swimming and driving for prey, all recorded for mainland *herodias* (Bent, 1926) and the Common Heron of the Old World (Lowe, 1954). This presumably reflects only lack of observations of unusual feeding methods.

Audubon (quoted in Bent, 1926) records that *herodias* pierces its prey with its bill, but I have never seen this or any species of heron secure food by stabbing. The prey is usually taken in scissor fashion; that is, it is grasped between the mandibles. Lowe makes similar statements for the Common Heron. Richard Borden (*pers. comm.*) informs me that he has a motion picture record of a Florida Bay *herodias* stabbing its prey.

If the prey proves to be too large for immediate consumption, *herodias* and other species will hold it firmly between the mandibles, then batter it against the ground, a rock, or a root, or they will repeatedly drop, strike, pick up, and drop it, until the prey is reduced to insensibility. Small prey is tossed directly into the gullet with one smooth movement of the head. Bill-dipping and head-shaking after swallowing prey are just as characteristic of *herodias* as of *virescens*, and no significant differences in these movements could be found.

Care of the Body Surface

Preening. I made no detailed observations of preening, but my notes suggest no striking differences between *herodias* and *virescens*. Only two points need to be mentioned: (1) *herodias* appeared to droop the wings more frequently and further when preening the upper neck (compare Plates 1 and 8); and (2) *herodias* only occasionally fanned the wing to any appreciable extent when underwing preening; normally birds merely extended the wing slightly and then put the head between the body and the partly open wing (compare Plate 2). Plate 9 shows a nestling white phase *herodias* preening one of its primaries.

Scratching. A. herodias scratches by bringing its leg up directly, the wing being held closed. The crest is always erected fully, and the head is rotated as the bird works the side of the head and neck with the middle toe.

Shaking. This series of movements is much like those of *virescens*, although it is my impression that tail-wagging is more often a terminal component of shaking in *herodias* than in *virescens*.

Bill-cleaning and other maintenance activities. Cleaning and wiping the bill, pecking the feet, and stretching the wings and legs showed no significant differences from the Green Heron. Plate 10 shows a nestling white heron in the sun-bathing position. Adult *herodias* show the same movements while sun-bathing, but I have never seen a sun-bathing *herodias* tilt the head upward as *virescens* often does.

The other maintenance activities (defectation, sleeping, yawning, and relaxation) differed in no significant aspects from *virescens*.

Alarm, "Freezing," and Escape Responses

Both phases of *herodias* show a variety of responses to danger. These may be grouped into two main categories: (1) flight; and (2) alert postures. I have never seen the Bittern Stance in *herodias*.

Flight. When fully aware of the approach of humans, *herodias* usually responds with an unhurried flight; such leisurely flights are usually silent. On occasion, however, a low, rather harsh "frahnk" or "frawnk" call may be uttered as the bird takes wing. When suddenly startled by a human, *herodias* explodes into flight, usually giving one or more "frawnk" calls.

These large herons vary a great deal in their responses to the close approach of a human to the nest. The incubating or brooding bird, if well aware of the intruder's presence, rises from the nest and stands in one of the alert postures described later. If suddenly surprised on the nest, however, these birds flush directly, and their subsequent behavior is highly variable. Some individuals land near the nest and keep the intruder in sight at all times, while others leave the area and circle about the nest site, sometimes landing hundreds of yards away. Others may fly back and forth over the nest, and each time they spot the intruder they fly rapidly away again. It is my distinct impression that those birds which fly around the nest call "frawnk" or "frahnk" more frequently and more intensely than those which flush and land nearby. As with the Green Heron I have never seen *herodias* hover

over the nest or the intruder, as is so characteristic of the Black-crowned Night Heron.

Alert postures. The measure of the escape tendency in any alert posture is the degree to which the neck is extended and the feathers are sleeked. In general, the greater the escape tendency, the greater the extension of the neck and the greater the sleeking of the plumage. In addition, the bird depresses the rear portion of its body more and more as it readies itself for flight. A bird in the normal perch posture is relaxed, with the head and neck withdrawn or only partly extended, and its feathers are neither erected nor sleeked.

My observations of *herodias* were not as complete as those of *virescens,* but crest-raising and tail-flicking are clearly *not* components of these alert postures in *herodias.*

An extreme escape response was released among white and blue phase birds which frequently fed in close proximity to a nest occupied by a pair of Ospreys, *Pandion haliaetus* (Linnaeus). Both adult Ospreys normally remained close to the nest, one incubating or brooding, the other guarding. Whenever a feeding heron approached too closely, the Osprey on guard would dive at the heron. As the Osprey swooped, the heron suddenly ducked down to an extremely low, flat position, head and neck withdrawn, feathers completely sleeked, the ventral parts of the body actually plunging into the water. As the Osprey veered away, the heron adopted an extreme alert posture, head and neck fully extended; in addition, the heron turned its body so as to keep the Osprey continually in full view. Occasionally, as the heron rose from the flattened position, it would utter one or more harsh "frawnk" calls. Only once did I observe a white phase *herodias* stab at one of the Ospreys as the latter swooped at the heron's head. The stab was made from the extreme crouched position.

Feeding *herodias* adopted moderately intense alert postures as they approached the vicinity of the Osprey's nest. Many times they bypassed the Osprey nest by feeding in roughly a circular path around the nest. The dives of the Ospreys were directed at the large herons (for example, *A. herodias, C. albus, D. rufescens*); the small species were ignored.

"Freezing" responses. Occasionally both phases of *herodias* will adopt an extreme alert posture, with head and neck elongated to the utmost, and bill horizontal. The birds freeze in this position for long periods. Sometimes the source of danger seems to be hidden, but the blue phase especially adopts this extreme alert posture when standing in reed-choked areas; these birds are fully aware of the nature of the disturbance. They seem to be trying to conceal themselves as much as possible. Stone (1937) gives a good description of these freezing postures in the blue phase of *herodias* in the marshes near Cape May, New Jersey. This posture is not to be confused with a Bittern Stance.

Other responses. I have never observed predation on *A. herodias.* The large size, powerful striking apparatus, and formidable threat displays of this species appear to be sufficient to intimidate most potential predators, but there are occasional references to adults being taken by predators (for example, Monson, 1951; Cottrille and Cottrille, 1958).

The Florida Bay populations of *herodias* were habituated to the roar of the numerous outboard motors, and other boat noises so common in this region, especially along the well-traveled intracoastal waterway. Away from the main boat lanes, however, these birds were very wary, and they usually responded to the sudden noise of a boat motor starting by showing an extreme alert posture or by flushing wildly.

Habituation to the click of a camera was quite rapid with one pair of *herodias* nesting on Cotton Key in Florida Bay (Meyerriecks and Meyerriecks, 1958). When the first migratory *herodias* arrived on Rulers Bar Hassock in 1957, the birds which fed near the tracks flushed wildly at the approach of the trains, but habituation was rapid. Later it was a common sight to see these and numerous smaller herons feeding within a few feet of the speeding trains.

My notes from the Florida Bay and Rulers Bar Hassock regions include many observations of the responses of *A. herodias* to the alarm notes of other birds. Table 16 lists the species and their calls which released alert postures or flight (unusual). Most passerine alarm notes were ignored, but occasionally a particularly loud note, given close to the heron, would elicit a very low intensity alert posture.

TABLE 16

BIRDS AND THEIR CALLS WHICH RELEASE ALERT POSTURES AND/OR FLIGHT IN *Ardea herodias.*

Species	Call
Redwinged Blackbird, *Agelaius phoeniceus*	"check"
Green Heron, *Butorides virescens*	"skeow"
Little Blue Heron, *Florida caerulea*	"aaah"
Reddish Egret, *Dichromanassa rufescens*	"crog"
Common Egret, *Casmerodius albus*	"frawnk"
Snowy Egret, *Leucophoyx thula*	"aarg"
Louisiana Heron, *Hydranassa tricolor*	"raah"
Black-crowned Night Heron, *Nycticorax nycticorax*	"squok"
Yellow-crowned Night Heron, *Nyctanassa violacea*	"squak"
American Bittern, *Botaurus lentiginosus*	"wok-wok"

MIGRATION AND BEHAVIOR ON ARRIVAL

Florida Bay *herodias* are non-migratory, but little is known about movements around their restricted range. At no time during my stay in Florida Bay (December 10, 1955–May 20, 1956) did I notice any changes which might indicate movements into or away from this area. Occupation of the breeding sites is protracted, and high winds and low temperatures inhibit courtship activities. The same applies to the blue birds in the Florida Bay area. Breeding may take place at any season of the year, but most eggs seem to be laid during the period September–March. On Cotton Key all eleven nests discovered in February, 1956, contained young; no nests had eggs. The young ranged from newly hatched birds to those capable of sustained flights (probably 50-60 days old). Courtship activities were noted around Cotton Key in March, however, so it is possible to observe at one time, even on one tiny key, birds in various stages of the reproductive cycle.

Many populations of mainland *herodias* are highly migratory. Stone (1937:113) suggests that male *herodias* breeding in southern New Jersey arrive before the females, but his description does not seem to support his conclusion. Detailed studies of the closely related *A. cinerea* in Europe reveal that both sexes arrive on the breeding grounds at the same time (Verwey, 1930; Lowe, 1954). The occurrence of dancing ground behavior (see p. 100 and Lowe *op. cit.*) suggests that there is a delay in the occupation of the breeding areas, but Roberts (1936:175) states that Great Blue Herons in Minnesota move directly into nest sites on arrival, and Cottrille and Cottrille (1958) show that courtship activities do not begin at once upon the return of the herons in spring to the Michigan colonies, even though they occupy the breeding sites on arrival.

This suggests behavioral differences between sedentary and migratory populations. Dancing ground behavior is indulged in by *herodias* in Florida Bay, but I have only been able to find two brief notes outside of the Florida Bay region (Mitchell, in Bent, 1926:102; Smith, 1894). Smith's observation was of sedentary populations in California. Both descriptions are too limited to be of use.

In the British Isles *A. cinerea* has been observed dancing prior to colony occupation (Lowe), but this behavior is unrecorded on the continent. The British populations are relatively sedentary, while the continental ones are highly migratory. There remains the distinct possibility that the migratory populations of both *herodias* and *cinerea* dance during migratory stops.

HOSTILE BEHAVIOR

Non-aerial, Non-vocal Displays

Crest-raising. The crest of the Green Heron is its most mobile feather unit, but the back, neck, and pectoral plumes of both mainland and Florida Bay *herodias* are much more mobile than the crest feathers. Crest-raising appears to be a component more of hostile than sexual displays in *herodias*.

The crest is fully erected in the Aggressive Upright, Forward, and

Full Forward Displays. The crest is not raised in any alert posture or in the Upright Display, both of which express high escape tendencies, or at least greater escape than attack tendencies. During the Snap Display the performing male erects its crest fully; however, I believe that this crest-raising is not a hostile component but functions to emphasize a sexual display.

A bristling blue phase bird, especially one in an intense hostile display (for example, the Forward Display), has the bill and the bulging yellow eyes accentuated by two triangular patches of black feathers on the sides of the head. These patches appear to point toward the eyes. This pattern is nicely shown for the similarly marked *A. cinerea* in Lowe (1954: Plate 4). The crest of *herodias* probably functions to increase the apparent size of the head of the attacking bird, thus accentuating the intimidatory function.

Tail-flipping. As with crest-raising, tail-flipping in this species is a component of those hostile displays which express greater attack than escape tendencies. I have seen tail-flipping only in the Aggressive Upright, Forward, and Full Forward Displays.

On occasion, but less frequently than with the Green Heron, a feeding *herodias* will flip its tail just before a strike. Perhaps tail-flipping is indicative of preparation for sudden movement at the sight of the prey.

Upright Display. A bird in this display (Figure 30) shows the following components:

1. the head and neck are extended upward to their fullest extent, and are held at about a 45 degree angle
2. the plumage is markedly sleeked, the neck especially appearing extremely thin
3. the bill remains closed, and the eyes seem to bulge
4. the wings are held tightly closed
5. no calls are given at anytime, and tail-flipping is absent.

On numerous occasions I saw both white and blue phase *herodias* perform mutual Upright Displays in the Florida Bay region. Such mutual threat displays afforded direct comparisons of the two phases, and in all cases the two forms were practically images of each other.

I believe that the escape tendency is relatively high. When contrasted with more hostile displays (Aggressive Upright or Forward, for example), the sleeked plumage, tightly closed wings, upright posture, and complete lack of calls indicate escape components. An increase in the attack tendency is expressed in the assumption of the Aggressive Upright Display, together with walking or wading toward the opponent. An increase in the escape tendency is shown by a turning to one side and slight extension of the wings, an intention movement of flight.

The Upright Display is most commonly seen where large numbers of *herodias* are feeding in close proximity, and assumption of this display usually indicates some sort of territorial trespass, or at least

FIGURE 30. Adult Great White Heron in the Upright Display.

FIGURE 31. Adult Great Blue Heron in the Aggressive Upright Display.

too close approach. Upright threats are also common during "dancing ground" activities (p. 100).

Aggressive Upright Display. This display differs from the preceding in the following respects: (1) the head and bill are arched downward by a shallow arching of the upper neck; (2) the crest, pectoral, back, and especially the neck feathers are all erected; this gives the neck a thick appearance; (3) very frequently the bird opens its bill and utters one or more harsh "rok-rok" calls; and (4) the bird moves toward the opponent in a stiff, tense fashion. Figure 31 shows a blue phase *herodias* in a typical Aggressive Upright Display, while Plate 4 in Lowe shows a striking similar, if not identical, display in *cinerea.*

Aggressive Uprights are much more frequently seen during feeding territory disputes, presumably because of the concentration of birds, but they are commonly seen, too, both in the breeding and non-breeding seasons, away from feeding areas.

The Aggressive Upright is an expression of a higher attack tendency than the Upright, and I believe that the attack tendency is relatively and actually much stronger than the escape tendency. The arched neck (preparation for a strike), feather erections, and harsh calls are all attack components, but the upright neck and tense attitude of the displaying bird strongly suggest conflict.

The Aggressive Upright Display of *herodias* resembles the Stiff-necked Upright Display of the Green Heron in the thick, stiff, upright attitude of the neck, and the downward inclination of the bill (compare Figures 10 and 31). The display of the Green Heron, however, occurs only under special circumstances during the breeding season, and only males perform this display. The Aggressive Upright of *herodias* may occur in any hostile clash, breeding season dispute or otherwise, and both sexes show this display.

Forward Display. From the Aggressive Upright, a bird assuming the Forward Display: (1) inclines the body forward to a horizontal position; (2) retracts its head and neck in preparation for a strike; (3) erects all of its feathers, especially the back plumes, to an extreme degree; (4) walks rapidly toward the opponent with its wings partly spread; (5) opens its bill wide and utters a series of rasping "rok-rok" calls; and (6) the bird's eyes appear to bulge from its head (Plate 11).

During the breeding season, when the soft-part colors change radically, an attacking *herodias* is a formidable creature, taking full advantage of its brilliant red-orange bill, bulging yellow eyes surrounded by lime-green lores and orbital skin, bristling plumage and bright red legs. However, the primary function of such soft-part color changes is to enhance sexual displays; their role in hostile displays seems secondary (see Part III).

The Forward is an expression of the strongest attack tendency shown

FIGURE 32. Great Blue Heron performing a Snap Display.

by *herodias,* short of the Full Forward which immediately precedes attack.

Full Forward Display. I doubt that this display is distinct because it is a forward strike with the head from the Forward posture. However, if the opponent retreats somewhat, the attacking bird will hold its head and neck in the extended position, which, as a definite modification of the Forward Display, is assigned a descriptive name.

The hostile series, Upright → Aggressive Upright → Forward → Full Forward, forms a sequence showing an increasing attack tendency and relative decrease in the escape tendency. In general, the displaying bird shows two major movements with a rising attack tendency: (1) increasing forward inclination of the body, and (2) increasing erection of plumage. The reverse movements are shown with an increasing tendency to escape; that is, the plumage is sleeked and the bird adopts an upright posture with head and neck extended fully.

Fighting. When fighting, *herodias* attempt to buffet their opponents with their wings and to strike them with their large bills. Although the initial thrust with the bill is typically made from the horizontal Full Forward position, subsequent strikes are delivered downward as each bird tries to get above the other. I have never seen a fight which resulted in damage to either of the combatants. Indeed, actual physical contacts are quite rare, most of the encounters being rather short-lived sparring matches.

Snap Display. Verwey's (1930:21) description and illustration of the *Schnappbewegung* of *A. cinerea* fit the same display in *A. herodias* exactly. With the exception of bobbing, which is a movement added to the Snap Display, I was unable to find any difference between *herodias* and *cinerea.*

The displaying bird, always a male as far as I was able to determine (see p. 33; the account of Cottrille and Cottrille is not at all clear on the Snap Display), typically performs while perched on the platform of an old nest. The male extends its head and neck fully out and downward until the bill is level with, or slightly below, the level of the feet. Then the male bends slightly at the "heel" joints, lowering the body toward the floor of the nest. Just as the head and neck reach their fullest extent the male snaps his mandibles together, producing a loud "bok" sound. The crest feathers, together with the occipital plumes, are fully erected during this display, but erection of the neck feathers is highly variable. After making the "bok" sound the male holds the extended pose for a few moments, then rises to a normal perched position (Figure 32).

One variant of the Snap Display, the Low Bow Display, is frequently shown by males which have been performing repeated Snaps. I believe that the Low Bow variant is a high intensity form of the typical Snap Display, but I lack quantitative data. The displaying bird assumes the extended posture of the Snap, except that all feathers remain depressed, no "bok" sound is made, and the neck is held in a shallow, inverted U-shape. In this position the bird bobs rapidly up and down, bending at the "heel" joints with each bob.

The Snap Displays, together with the bobbing, of *herodias* and *virescens* are strikingly similar. The similarity is heightened by the fact than an occasional male *herodias* will grasp a twig momentarily at the end of a Snap Display, just as male *virescens* grasp, then release, twigs after a Snap. My observations of the Snap in *herodias* were less detailed than those of *virescens,* but I believe that the causation of this display is similar in both species (see discussion in Part I).

Lowe describes a variant of the Snap Display in *cinerea* which appears to be entirely hostile in its causation. This hostile Snap apparently occurs only during territorial disputes, but the form of the display is identical with the Snap shown by the male toward a sexually active female. I have not seen this hostile form of the Snap in *herodias,* and no mention of such a display by *herodias* is made by Cottrille and Cottrille.

Bill-snapping. Verwey has described this behavior for *cinerea,* but I have not seen it in *herodias.* However, bill-snapping and several other important pre-copulatory displays are described by Cottrille and Cottrille. These displays in *herodias* resemble closely similar displays in *cinerea* (Lowe).

Withdrawn Crouch. The only behavior of *herodias* that resembles the Withdrawn Crouch of *virescens* is the extreme crouch described on page 91. It is possible, however, that something similar to the Withdrawn Crouch of the Green Heron might be observed in a large colony of *herodias* where birds are continually landing in the crowded nesting areas.

Dancing ground displays. This behavior has been referred to briefly, and Lowe gives a good account of similar activities in English populations of *cinerea*.

I have seen this behavior only in the Florida Bay region, but casual notes in the literature appear to deal with dancing ground behavior in mainland *herodias* (Smith, 1894; and Mitchell, in Bent, 1926). Burleigh (1929) describes an observation made near Renton, Washington, on March 14, of a circling, tumbling flight in *herodias* which might be an aerial prelude to dancing ground activities. The flight resembled those described by Lowe for English *cinerea*.

An observation I made near Bottlepoint Key, eastern Florida Bay, on December 26, 1955, is typical of dancing ground behavior. Four white and one blue *herodias* were standing quietly in the shallow water immediately south of Bottlepoint Key. One of the whites and the blue had brilliant soft-part coloration, and these two appeared to start the dancing activities. The brilliantly colored white bird raised its head and neck to about a 45 degree angle, erected its back plumes, and strutted toward one of the other white birds. Immediately, the blue bird performed in the same fashion, but the display of the blue bird was abruptly stopped when the bird (dull white) toward which the display was directed threatened the displaying blue. Both the latter and the dull white bird assumed the Upright Display posture, which they held for almost one minute; then both threatening birds relaxed to an extended alert posture. At no time did the two birds that danced display toward each other, and, because of their dancing, soft-part coloration, and subsequent behavior, I took them to be males.

A short time later, the blue bird extended its wings very slightly, erected its head and neck as before, and danced toward a dull-colored white bird. The dance was a combination of a skip and a slight jump-flight, and the blue bird seemed to float momentarily. After it had gone about ten yards, the blue bird took off and circled around the white bird toward which its dance had been directed, and its circular flight reminded me of the Circle Flight Display of the Green Heron.

On two other occasions I saw this dancing ground behavior in the Florida Bay region, and Robert P. Allen, of Tavernier, Florida, informs me (*in litt.*) that he has seen this performance many times in the same region. Both Allen and I have seen strutting and dancing in birds that ranged in soft-part coloration from brilliant to very dull (that is, non-breeding coloration). Two facts, the dullness of the soft parts of some "dancers" and the frequency of Upright Displays during the performance, lead me to conclude that dancing ground activities are the first indications of breeding season activities in *A. herodias*, at least in the Florida Bay area. Although more observations are necessary, I believe that the upright posture and strutting walk of the dancers has evolved from the Upright Display; this is suggested by the extended head and neck, sleeked plumage (except for the back plumes), and orientation of the dancing display. Hostile displays are common precursors of many sexual displays in birds.

Non-aerial, Vocal Displays

Advertising calls. Verwey (1930) and Witherby *et al.*, (1949) record a pairing or advertising call for *A. cinerea.* I have no information on such calls for *A. herodias,* although Cottrille and Cottrille state, "this eerie howl of the heron is perhaps its nearest counterpart to the territory-advertisement and mating songs characteristic of many birds." The eerie howl is the call that accompanies the Stretch Display. They list the calls of *cinerea* and state, "those of the Great Blue Heron [are] very similar," but no details are given.

On several occasions in Florida Bay, I observed a male, standing on an old nest, open and close his bill repeatedly, but I was unable to approach close enough to establish the nature of the call. Judging from the behavior of the Green Heron and other heron species, an advertising call was appropriate for the stage of the breeding cycle and the location of this male.

Attack call. Other than some harsh croaks, difficult to describe, heard during a fight, the only attack call I have ever heard in *herodias* is the harsh "rok-rok" (p. 97).

Verwey describes several hostile calls for *cinerea,* among them a loud "gooo" uttered by a bird as it stabs at an opponent, and a harsh "schaah" uttered by the birds in a Pursuit Flight. Cottrille and Cottrille state that similar calls are made by *herodias.*

Escape and alarm calls. The only notes associated with escape and alarm that I heard were the "frawnk" and "frahnk" calls (p. 90). When merely aware of the presence of a source of threat, the alarmed bird utters one or more barely audible, low "frawnks." As the escape tendency increases, however, the "frawnks" are louder, harsher, and more frequent.

Cottrille and Cottrille describe a cry of "pain or terror" (*herodias*) heard on one occasion, and Verwey describes an "anxiety" note for *cinerea* (a nasal "go-go-go"). This is uttered especially when a disturbance, such as a man near a nest, is located below the calling bird (Witherby *et al.*).

Aerial, Non-vocal Displays

I observed no Flying Around, Pursuit Flights, Supplanting Attacks, or ritualized display flights by *herodias.* The evidence in the literature suggests that there is little or no aerial behavior preceding territorial activities. Cottrille and Cottrille mention a pursuit call "made by both pursued and pursuer." One observation by the Cottrilles (1958:5) suggests a mutual aerial display, but no details are given. The circular flight which I observed on December 26, 1955 (p. 100) indicates that a more detailed study might reveal aerial displays in *herodias*. It is my impression that few aerial displays are shown by the larger species (e.g., *A. herodias, A. cinerea, C. albus*), while aerial displays are very common with the smaller species (e.g., *B. virescens, L. thula*). I observed no aerial, vocal displays by *herodias*.

SEXUAL BEHAVIOR

I observed only one sexual display, the Stretch Display, in the Florida Bay region, but Cottrille and Cottrille describe a number from Michigan.

The Stretch Display is the same as the *Reckbewegung* of *Ardea cinerea*, so well described and figured by Verwey, Holstein (1927), Witherby *et al.*, and Lowe. Cottrille and Cottrille, who call what is obviously a Stretch Display "howling," make no comparison with the same display in *cinerea*, although their figures beautifully show that the display they observed was the Stretch.

I had Verwey's description and illustration before me in the field, and I was unable to detect the slightest difference between the displays of the two species. I was, however, unable to hear clearly the call which accompanied the Stretch Display of *herodias*, but the description of this call given by Cottrille and Cottrille shows that the calls of *herodias* and *cinerea* are strikingly similar. Verwey and Lowe state that males of *cinerea* utter a two-parted "hu-oooo" or "hoo-oooo," and Cottrille and Cottrille state that *herodias* emits a "low o-o-o-o"; at the height of the Stretch the sound becomes higher and louder—"h-o-o-o-o-o."

The males I saw performing the Stretch Display did so from the floor or rim of an old nest, less frequently from a nearby branch; similar observations were made by the Cottrilles in Michigan. When displaying the Stretch, the males I saw made the following movements:

1. the bird stretched its head and neck to their fullest, bill pointing straight up
2. it depressed all of its feathers except the pectoral plumes
3. it held this position momentarily, then lowered its head and neck down and backward so that the occipital plumes almost touched the middle of the back
4. as the bird lowered its head and neck, it bent its legs at the "heel" joints and lowered its body toward the nest
5. it held this lowered position for a few moments, then rose to a normal standing position (Figure 33).

All the males of *herodias* I saw perform the Stretch Display in the Florida Bay area had brilliant soft-part coloration. The description

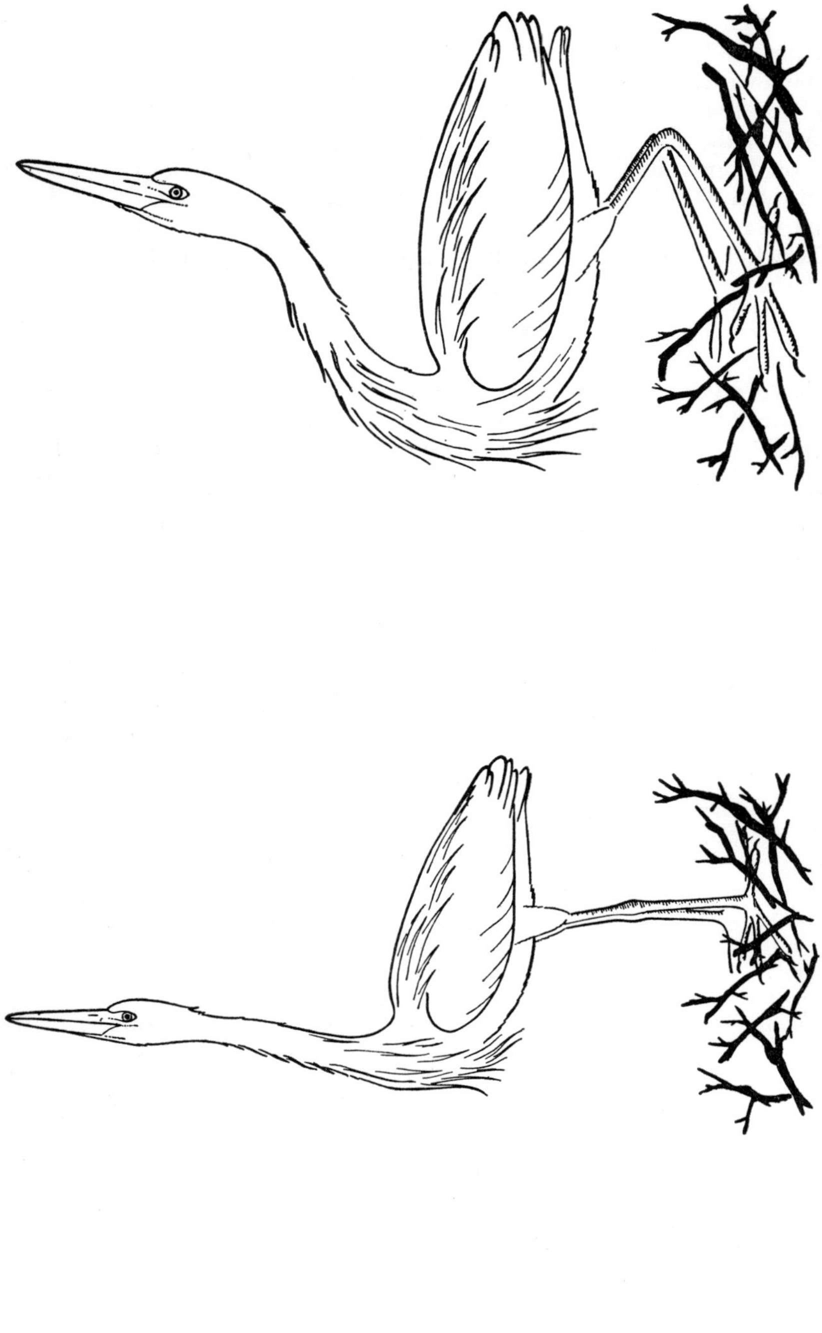

FIGURE 33. Great White Heron performing the Stretch Display. The figure to the left represents stage 1 of the Stretch, while stage 2 is shown in the figure on the right.

given by the Cottrilles for Michigan *herodias* is similar, except that the birds they saw appeared to erect their neck feathers to a great extent, and when the displaying bird arose from the crouch, it held its bill pointing upward; furthermore, no remarks are made by the Cottrilles on the coloration of the birds they observed.

I believe that the Stretch of *herodias* is simpler and more highly ritualized than the same display of *virescens*. Nothing that I noted resembled the Extended Stretch Display of *virescens*. I get the impression from the descriptions given by Cottrille and Cottrille that the Stretch Display ("howling") of Michigan *herodias* is primarily sexual in nature.

I conclude that the origin and function of the Stretch Displays of *virescens* and *herodias* are similar.

EARLY PAIRING BEHAVIOR

I have no information on pairing behavior in *herodias* after the initial occupation of the breeding areas. Details on many aspects of pairing biology of this species are to be found in Cottrille and Cottrille.

No detailed studies of roosting behavior were made, but *herodias* usually alights in the crowns of the trees, then walks or jumps slowly down to the final roosting spot, while *virescens* characteristically flies directly down to the final spot, or jumps down to it.

A vast literature, mainly anecdotal, on coloration of the soft-parts of herons has accumulated, but little pertains to North American species. Most of the data on soft-part coloration I gathered in the Florida Bay region, but I also made some important observations of

TABLE 17

SOFT-PART COLORATION OF *Ardea herodias* AND *A. cinerea*.

Area	*A. herodias* (white phase)	*A. herodias* (blue phase)	*A. cinerea*
		Non-breeding	
bill	dull yellow	dull yellow	yellow
lores and orbital skin	pale grayish-blue	pale grayish-blue	yellowish-green
legs and feet	yellowish-green	brownish or greenish black	brownish
iris	yellow	yellow	yellow
		Breeding	
bill	bright orange to red	bright orange to red	orange to red
lores and orbital skin	bright lime-green	bright lime-green	red ? ?
legs and feet	bright red	bright red	pinkish to red
iris	yellow	yellow	yellow

migratory *herodias* on Rulers Bar Hassock in the spring of 1957. No color changes were noted by Cottrille and Cottrille in Michigan.

Table 17 lists the breeding and non-breeding coloration of the soft-parts of both color phases of *herodias* and, for comparative purposes, *A. cinerea*. I have not listed sexual differences because of insufficient observations. However, *A. herodias* males are more brilliantly colored than the females and acquire their brilliant breeding coloration earlier than the females.

REDDISH EGRET

Bock (1956), on the basis of a few plume characters, places *Dichromanassa rufescens, Hydranassa tricolor,* and *Florida caerulea* in one genus (*Hydranassa*), and he puts *Leucophoyx thula* in another (*Egretta*). My observations of the behavior of *rufescens* and *thula* indicate that they are congeneric. Bock admits, "The lack of characters distinguishing *Hydranassa* from *Egretta* presents a major problem. Other than the difference in plumes, I have been unable to find any good character that separates the two genera."

METHODS

With the exception of a few minor observations near Puerto Arista, Chiapas, Mexico, I made all of my observations of the Reddish Egret, *Dichromanassa rufescens,* around the Bottlepoint, Stake, Low, and Cowpens Keys area of Florida Bay (see map in Meyerriecks, 1957b). I used the same techniques as with the previous species. However, *rufescens* is an extremely tame species, and I was able to make numerous observations from an open boat anchored in the shallows. The areas have already been described.

DISTRIBUTION AND DESCRIPTION

The Reddish Egret ranges as a breeding bird from southern Florida, Louisiana, the coast of Texas, and Baja California, south to Sinaloa, Yucatan, Cuba, Isle of Pines, and the Bahamas. The species winters south to Guatemala, El Salvador, and Venezuela. A more detailed treatment of the breeding range and frequency of the white and dark phases is to be found in Allen (1954–55). The Reddish Egret nests singly or in small to large colonies. The largest breeding colony known (several thousand pairs) is on Green Island, in the Laguna Madre, off the coast of Texas. At all seasons of the year this bird is almost completely confined to salt or brackish water.

D. rufescens is medium-sized, length about thirty inches, wing spread about four feet. Most field guides state that it occurs in two color phases, light and dark, but this is misleading because the Reddish Egret is extremely variable in color and pattern, at least in the dark phases. My field notes clearly indicate that there are at least three color phases (white, intermediate, dark), a statement which Bock also makes, although he gives no details. Allen (1954–55) also recognizes three phases. I have numerous notes on highly aberrant forms from the Florida Bay region, including observations on two melanistic birds. These notes suggest even greater complexity than three phases, but this information will be published elsewhere.

Adults have numerous lanceolate plumes covering the entire head and neck, these plumes being especially long in the occipital and pectoral regions. Long scapular plumes extend beyond the tip of the tail in all phases. The dark birds (except the two melanics noted) have the head and neck colored a vinaceous-brown, with the body and wings neutral gray. The intermediate birds are neutral gray in color all over, and the white phase birds are typically immaculate; some white phase individuals, however, may have a few to many of the flight feathers tipped with slate gray. The basal half of the bill in adults is a pale flesh color (non-breeding season), while the tip is black. All phases have blue legs and feet.

MAINTENANCE ACTIVITIES

Locomotion. No detailed studies of take-off and landing were made, but *rufescens* does not appear to differ significantly from the previous species in these activities. Except in hostile situations the Reddish Egret does not erect its crest or flip its tail conspicuously while landing or taking off.

The flight is strong and rapid. Except for some unusual aerial displays to be described later, the flight of *rufescens* is direct. On long flights the head and neck are retracted, but on short flights, and in certain specialized displays, the head and neck are extended.

Unlike *herodias* and *albus, rufescens* is extremely active. In the grace and agility of its movements the Reddish Egret is matched only by the Snowy Egret and the Louisiana Heron, among North American herons.

This species has a very striking spiral descent flight which most closely resembles that of the Snowy Egret and the Louisiana Heron. Spiral descent flights are extremely common during the courtship period.

Green Herons walk by sneaking forward from a crouched position; *herodias* and *albus* wade forward smoothly and unhurriedly from an upright position; *rufescens* strides along with rather short, jerky motions from a peculiarly upright position. No other heron with which I am familiar walks in the same manner.

Feeding behavior. The Reddish Egret is unique among North American herons in its varied feeding behavior. Not only does it show the previously described Stand and Wait, Wade or Walk Slowly, Peering Over, and Wing-flicking types, but it possesses an almost incredible number of other types. These other types all involve a great deal of activity, and all (with the exception of Canopy Feeding) are variations of the general category called Disturb and Chase by Meinertzhagen (1949), who used this term with reference to the feeding behavior of the Reef Heron, *Demigretta gularis* (Bosc).

The basic types, Stand and Wait and Wade or Walk Slowly, are shown rather infrequently. Wing-flicking, in which the feeding bird moves forward slowly while partly extending and retracting its wings rapidly, is quite common, but the most characteristic feeding technique used by *rufescens* is a lurching, weaving type of half-run, half-jump progression. As the bird reels forward, it stabs rapidly to the right and left, attempting to seize any prey disturbed by its activities. This method is used primarily in very *shallow* water. One characteristic variant of Wade or Walk Slowly might be termed "head-tilting." The bird moves rather slowly through the water with its head and neck strongly inclined to one side as it peers intently at the surface (see Plate 12). If a school of tiny fish is disturbed, the egret chases the prey over a wide area, and it holds its head and neck tilted to one side or the other throughout the chase.

Open Wing Feeding is another characteristic feeding technique of *rufescens*. The feeding bird begins by running slowly, wings partly extended. Then at the sight of the prey it runs rapidly and extends the wings fully. As the prey dash about, the egret turns and twists, wings still extended fully, and the bizarre performance may continue for several minutes. On occasion the bird may halt suddenly in the middle of a run, retract one wing, rapidly extend and retract the other wing, and then resume its forward run. Open Wing Feeding frequently ends in Canopy Feeding or Underwing Feeding.

Canopy Feeding has been described for the Black Heron, *Melanophoyx ardesiaca* (Wagler), of tropical Africa and Madagascar, by numerous authors (for example, Ayres, 1878; Loveridge, 1922; Rand, 1936; Jackson, 1938; Delacour, 1946; Winterbottom, 1957). Superb photographs of this unusual feeding technique are to be found in Delacour (1946) and Winterbottom (1957). To my knowledge this type of feeding behavior has not heretofore been described for the Reddish Egret. In fact, Delacour, speaking of *ardesiaca*, states, "it has a very peculiar way of catching fish, which I think is unique in the family." Although *ardesiaca* shows Canopy Feeding to an extreme degree, it still closely resembles the feeding of *rufescens*.

When showing the Canopy Feeding method, *rufescens* runs forward with wings extended, then halts and peers into the water, and then brings both wings forward over its head, forming a canopy over the head and neck. This pose is held for a few moments to several minutes. I could clearly see the rapid fish-catching movements the bird made under the canopy of the wings. This species differs from *ardesiaca* in that the latter typically places the tips of the wings in the water, while all the Reddish Egrets I saw Canopy Feeding held the tips of the wings clear of the surface. Plates 13 and 14 show two expressions of Canopy Feeding by the Reddish Egret.

Rand (1936), describing the canopy of *ardesiaca*, concluded that the function was to bewilder the prey by the darkness of the canopy. My observations of *rufescens* at extremely close range (three feet) indicate that the shadow provides a false refuge for the fish startled into motion by the previous dashing activities. Reddish Egrets typically hold their wings in the canopy attitude for a minute or two *before* they make a strike. I could clearly see many fish enter the shade of the canopy in the shallow water, and when a number of prey had thus "fallen for the ruse," the egret would stab rapidly under its extended wings.

Very often Reddish Egrets engage in feeding which I believe is low intensity Canopy Feeding. I call it Underwing Feeding. The egret will run forward, wings fully extended, then stop suddenly, extend one or both wings fully and horizontally, then stab rapidly under one or both wings. This type differs from Canopy Feeding in that the wings are not brought forward and the bird does not place its head under the raised wings except when it strikes.

I conclude that the following series of feeding types represents an evolutionary sequence: Wing-flicking → Open Wing → Underwing → Canopy Feeding. The Great Blue Heron, Common Egret, Reddish Egret, Snowy Egret, Louisiana Heron, and the Little Blue Heron all show the initial stage, Wing-flicking. Sutton (1936) gives an excellent description of what is obviously Wing-flicking feeding behavior in the Least Bittern, *Ixobrychus exilis* (Gmelin). The Reddish Egret, Snowy Egret, and the Louisiana Heron show further developments of the series, but the Reddish Egret shows *all* the steps. From the descriptions of Canopy Feeding by the Black Heron, it appears that this species has carried the evolutionary series to its extreme development. It would be of the greatest interest and importance to discover what other heron species show any or all of these stages.

Although there are minor variations in all of these feeding types, only two other methods warrant description. These are foot-stirring and hovering-stirring. In a recent paper by Rand (1956) the Reddish Egret was not listed as one of the North American herons that show foot-stirring. However, N. B. Moore (in Baird *et al.*, 1884) gives an excellent description of such behavior in *rufesceens.* I saw foot-stirring by the Reddish Egret on numerous occasions in the Florida Bay region.

As the egret waded forward rather slowly, it would vibrate its feet over the surface of the mud, imparting a scraping motion to the feet; then it would stop and peer at the surface of the water and either strike at prey or move on, usually resuming the scraping motions. Hovering-stirring is an aerial variant of typical stirring or scraping. The feeding bird moves forward slowly, then suddenly launches into flight, hovers over the surface of the water, and very gracefully scrapes the mud or aquatic vegetation with one foot. This behavior is repeated while the bird continues in flight. The strike is made from the hovering position. A more detailed treatment of foot-stirring and hovering-stirring by *rufescens* and other heron species is to be found in Meyerriecks (1959).

This brief treatment of the amazing array of feeding activities of *rufescens* emphasizes my statement that the Reddish Egret is the most active of the North American herons.

Care of the body surface. I made detailed observations on most of the other major maintenance activities of *rufescens* (for example, preening, head-scratching, and sunning), but this species does not differ significantly from the species already described. The Reddish Egret does, however, engage in underwing preening more frequently and more extensively than any other heron species with which I am familiar; in addition, such preening is most typical after very active feeding behavior. I believe that the greater frequency of underwing preening and the active feeding methods are correlated.

Alarm, "freezing," and escape responses. In its alarm and escape responses the Reddish Egret does not differ significantly from other heron species. Crest-raising in this species, as with *herodias* and *albus,* is not a typical component.

When caused to fly by the sudden appearance of a human, the Reddish Egret may utter a not unpleasant "crog" call note; on other occasions the startled bird may utter a rather harsh "raaah" note. Griscom (1926), after flushing Reddish Egrets from their nests on Culebra Key, Ascension Bay, Quintana Roo, recorded a nasal, croaking "wonk" call note.

The extension of the head and neck, together with thinness of the neck, are keys to the strength of the escape tendency in any alert posture. I saw nothing resembling a Bittern Stance in the Reddish Egret.

MIGRATION AND BEHAVIOR ON ARRIVAL

In Florida this species is migratory, but in many parts of its range it appears to be sedentary. I arrived in the Florida Bay area on December 10, 1955, and a few Reddish Egrets were already present. A definite movement into the Keys was noted on December 24, 1955, and the first signs of breeding activity were also noted the same day. Although I made no daily counts, I feel that the movement into the area was protracted. Allen (1954–55) states, for the Green Island population, "Usually the first birds arrive at Green Island from somewhere in Mexico in March, but most of them show up in April. There is considerable spread to these arrivals."

High winds and low temperatures definitely inhibited breeding activities in the Florida Bay region, and Allen (1942, 1954–55) noted the marked effect of temperature changes on the breeding behavior of *rufescens* off the coast of Texas.

HOSTILE BEHAVIOR

Non-aerial, Non-vocal Displays

Crest-raising. Of all the North American herons the Reddish Egret is easily the "bristliest," and this "bristling" is particularly striking in crest-raising. Moderate to extreme crest-erection occurs in a variety of hostile and sexual displays.

The non-aerial displays in which the crest is erected are the Upright (moderate erection), Aggressive Upright (full erection), and Forward (extreme erection) Displays. The degree of crest-raising is an excellent indicator of the strength of the attack tendency.

The Reddish Egret resembles the Green Heron in that the crest is its most mobile feather unit. When *rufescens* erects its crest, it appears strikingly "big-headed." Although crest-raising certainly functions to increase the apparent size of the weapon-bearing part of an attacking bird, I believe that hostility may be a secondary function. The extreme development of plumes and their erection have evolved chiefly to emphasize certain sexual displays in which head-tossing is a component (p. 118).

Tail-flipping. The only hostile displays in which the tail is flipped are the Upright and Snap Displays. On several occasions I saw a feeding Reddish Egret flip its tail just before a strike, but this movement was so rare that I suspect it is not a typical component of the strike as it is with the Green Heron.

Upright Display. A Reddish Egret in this display makes the following movements:

1. extends the head and neck upward to their fullest, inclines the head and upper part of the neck to the rear so that the occipital region is directly over the middle of its back—the bird is leaning backwards
2. holds the bill closed and horizontal
3. sleeks the body plumage to an extreme, but erects the crest and neck feathers moderately
4. holds the wings tightly closed, and may or may not flip the tail infrequently
5. gives no calls at any time
6. bulges its eyes from the head (Figure 34).

It should be evident that the Upright of *rufescens* differs significantly from the similar display of *herodias,* which shows no crest and neck feather erections, and inclines the body and extended head and neck *forward* to about a 45 degree angle.

I conclude that the escape tendency of a Reddish Egret showing this display is strong because a slight increase in the escape tendency almost invariably leads to immediate flight. Orientation away from the opponent, slight extension of the wings, and tail-flipping all indicate a rise in the escape tendency.

Aggressive Upright Display. This display, which shows a relatively greater attack than escape tendency, has the following components:

1. the bird holds its head and neck as in the Upright, but inclines them forward so that the occipital region is above the forward part of the back
2. holds its bill horizontal but opened, and the bird may utter one or more rather harsh "raaah" calls

FIGURE 34. The Upright Display of the Reddish Egret.

FIGURE 35. The Aggressive Upright Display of the Reddish Egret.

3. erects its body plumage, especially scapular plumes, moderately; erects crest, neck, and pectoral plumes strongly
4. holds wings partly extended
5. faces opponent frontally and walks slowly toward him
6. bulges its eyes from the head (Figure 35).

Orientation toward the opponent, wing extension, harsh calls, full feather erections, and forward inclination of the head and neck indicate a rise in the attack tendency. If the escape tendency of a bird in the Aggressive Upright increases, neck, pectoral and scapular plumes are depressed, and the head and neck are inclined rearward.

Forward Display. This display, which is the most intense hostile behavior shown by *rufescens* short of actual fighting, differs from the previous display in that the feathers of the head, neck, pectoral, and back regions are all erected to an extreme. Furthermore, the displaying bird extends its wings fully and *runs* directly toward the opponent, uttering a series of very harsh "raaah" calls as it rushes forward. A bird in the Forward Display fairly bristles all over, and the headlong rush toward the opponent is quite spectacular. The only conceivable tendency being expressed in this display is intense attack.

Fighting. I saw fighting on many occasions in the Florida Bay region. All of the fights were feeding territory disputes or clashes at the start of pair formation. Reddish Egret fights are truly spectacular, frequently ranging over a hundred yards or more. Mutual Forward Displays lead to headlong crashes. Both participants attempt to strike at the opponent from above; at the same time they jab viciously with the bill and flail with buffeting blows from the wildly flapping wings. Almost all of the blows made with the bill are aimed at the opponent's head. The Reddish Egret differs from fighting Green Herons in that the latter species strikes and recoils almost always from a horizontal body position, while *rufescens* fights from an upright position. The only damage I ever saw was the loss of a few feathers, but the force and tenacity of their fighting indicates real damage may be expected.

The hostile series, Upright → Aggressive Upright → Forward Displays, forms a sequence showing increasing attack tendencies. In general, as the attack tendency rises, the bird increasingly inclines its head and neck forward and increasingly erects its crest and other plumage.

Snap Display. This display in *rufescens* differs from the same display in *virescens* and *herodias* in the following components:

1. instead of a simple forward and downward extension of the head and neck, *rufescens* bows and extends at the same time, first to the right, then to the midline, then to the left, and then repeats
2. the bird erects the crest and neck feathers fully as it bows forward
3. *rufescens* shows tail-flipping as a typical component.

The Snap Display of the Reddish Egret appears to be more specialized in that only the bowing type is shown, while *virescens* and *herodias* show both the bowing type and the simple forward Snap without any bowing components. Furthermore, *virescens* and *herodias* typically erect only the crown feathers, while *rufescens* erects the neck feathers in addition.

The causation of the Snap Display of the Reddish Egret appears to differ significantly from that of the other two species. The Snap of *rufescens* is much more of a hostile display because: (1) the bird shows the Snap during the very early stages of pair formation; (2) the Snap is preceded and/or followed by other hostile displays; and (3) any individual approaching a bird giving the Snap is threatened and/or attacked. In contrast, *virescens* and *herodias* frequently alternate Stretch and other sexual displays with their Snaps, and the approach of another individual frequently triggers a bout of intense sexual displays. Study of a large population of *rufescens* (for example, Green Island, Texas), however, might reveal the occurrence of a Snap Display with greater sexual tendencies. It would be of great interest to determine the degree of tail-flipping, if any, in such sexual Snaps.

Dancing ground displays. On two occasions in the Florida Bay region I saw what appeared to be a dancing ground display by the Reddish Egret. Audubon (in Bent, 1926) describes what may be dancing ground behavior. In my first observation of this behavior six birds were feeding in one group on a large shoal. Two of the birds had brilliant soft-part coloration. A seventh Reddish Egret flew in from a nearby shoal and landed about fifty yards from the group. This new arrival also had brilliant soft-parts. Just before landing, "seven" raised his head and neck to about a 45 degree angle and began low intensity head-tossing. When head-tossing, the displaying bird rapidly raises and lowers its head in a vertical plane and fully erects the crown plumes.

"Seven" continued head-tossing after it had landed, and then it slowly moved toward the group. As it approached one of the dull-colored birds (a female?), "seven" opened its wings slightly, raised its head and neck upward and forward so that the head and bill were held at about a 45 degree angle, and then it began very intense head-tossing. At the same time "seven" began to utter a series of low, not unpleasant "crog-crog" calls. However, the displaying of "seven" was abruptly ended when one of the other brilliantly colored birds (a male?) threatened "seven" with the Upright Display. "Seven" immediately responded by adopting an Aggressive Upright; the two birds flew at each other and engaged in a brief fight. "Seven" succeeded in driving the other (male?) from the shoal, but I saw no further dancing.

The only other observation of dancing ground displays I made involved five birds feeding together on a small shoal. One of the birds (a male?), with very brilliant soft-part colors, started the dancing activities. This bird paused in its feeding, gazed briefly at one of the other Reddish Egrets, then raised its head and neck, extended its wings, and began to head-toss rapidly. The "actor," still head-tossing, moved slowly toward the egret it had been looking at, and then suddenly flew and began a most bizarre aerial performance. It landed, still head-tossing, on one side of the standing egret, then flew up and over the standing bird, landed on the other side, and then repeated the whole performance. The standing bird assumed a low intensity Upright posture each time the displaying male (?) flew over its head. This jumping over display was continued for a few minutes, then the "actor" landed and began to walk toward one of the other birds. Its walk was more of a ritualized strut as it moved very slowly toward the other bird. The dancing ground display was ended when one of the other egrets threatened the "actor." Mutual threat displays followed, then all five birds resumed their feeding.

Such dancing ground displays were the first indications of breeding behavior. My preliminary belief is that they are essentially hostile in nature.

Non-aerial, Vocal Displays

On several occasions I heard a bird, presumably a male, utter a series of "arg-arg" calls from the top of a large mangrove tree. These birds were always brilliantly colored, and all of them had previously shown rather intense Snap Displays. Other individuals, presumably females, were attracted to the vicinity of the calling males. Furthermore, the "arg" call of *rufescens* very closely resembles the "aarh" call of the Snowy Egret, which is the advertising call of that species. This "arg" call of *rufescens* may be the advertising call of this species, but I do not have enough data for proof.

Aerial, Non-vocal Displays

Flying Around. This behavior, one of the first indications of territoriality, is very similar to that of the Green Heron. I saw individual Reddish Egrets Flying Around repeatedly in the Cowpens Keys area of Florida Bay. They flew from one key to another, around the periphery of one key, or typically, back and forth from one end of a key to the other. All of the birds engaged in Flying Around were brilliantly colored, and all were especially responsive to the presence of other Reddish Egrets. I conclude that the function of Flying Around in *rufescens* is the same as in the Green Heron—to familiarize the bird with potential nesting sites and with those birds which are ready to engage in breeding activities.

Pursuit Flight. Once individual males had established themselves on one part of one key, they engaged in Pursuit Flights with any other Reddish Egrets which approached their territory. Most of the Pursuit Flights at the start of the breeding season were short-lived, mainly because the pursued did not appear to be ready to engage in any further activities. These birds were typically dull-colored. When the pursuer engaged a brilliantly colored bird in a Pursuit, however, the flight was usually quite prolonged and ranged over the entire length of one key, or from one key to another. Crest-raising by both pursuer and pursued was typical, but the crest-raising of the pursuer was more intense. Another characteristic component of lengthy Pursuit Flights was bill-snapping.

Supplanting Attack. As with the Green Heron the Supplanting Attacks of the Reddish Egret were most noticeable during the early phases of the reproductive cycle. The frequency and intensity fell off sharply once the pair began to engage in mutual sexual displays. The only difference between the Supplanting Attacks of *rufescens* and those of *virescens* is that *rufescens*, when attacking, erects all of its plumage as it supplants.

One variant of the Supplanting Attack of the Reddish Egret involves head-tossing by the supplanting bird during the short flight toward the bird to be supplanted. The attacking bird launches from its perch, then about midway through its flight it elevates its head and neck slightly and shows very low intensity head-tossing. I noted that in almost every case in which a Supplanting Attack was preceded by head-tossing, the attack was mild, and I suspect that the attacking bird was expressing a very low sexual tendency in its head-tossing.

Bill-snapping. This behavior is most commonly seen during prolonged Pursuit Flights, and the pursuer is the only individual observed to bill-snap. As the pursuer approaches the pursued, the former opens and shuts its bill rapidly, producing an audible snapping noise; this snapping may be repeated over and over during one Pursuit Flight, but it is typically shown on close approach. The usual response of the pursued is to elevate the crest strongly, and to turn and peer intently at the pursuer.

I believe that bill-snapping in Pursuit Flights is strictly a hostile component, because many of the birds which showed this aerial performance attacked the pursued when the birds landed, and they frequently attempted to nip the rear portions of the body of the pursued during the flight.

Aerial, Vocal Displays

On many occasions I heard individual Reddish Egrets utter the "raaah," "crog," and "arg" calls during flight in what appeared to be hostile activities, but I did not observe any aerial, vocal displays which appeared to be ritualized to any extent.

SEXUAL BEHAVIOR

The sexual displays of the Reddish Egret are given with striking frequency in both aerial and non-aerial situations. This is especially true of the Stretch Display. Another striking characteristic is the frequency of *mutual* sexual displays.

Stretch Displays. There are three distinct Stretch Displays: (1) a Stationary Stretch Display; (2) an Aerial Stretch Display; and (3) a Circle Stretch Display. Many low intensity forms of all three occur.

Stationary Stretch Display. The bird performs this display while perched in an exposed situation, and it makes the following movements:

1. it extends the head and neck fully upward, then brings the head slightly back and down so that the occipital region is over the upper part of the back—this is done in one smooth motion
2. it points the bill upwards to the zenith
3. it erects the feathers of the crown, neck, pectoral, and scapular regions to their fullest
4. it bulges its eyes from the head
5. just as the bill points to the zenith, the bird performs an intense bout of head-tossing, the head being moved rapidly up and down in a vertical plane (Figure 36).

I was unable to determine whether or not any call accompanies the Stationary Stretch Display, but the displaying birds kept their bills closed during the performance. A soft "crog" call, however, is a component of the other Stretch Displays.

In low intensity displays the bird holds the head slightly from the vertical and reduces the number and amplitude of the head-tosses. In some low intensity forms the head-tossing is eliminated. The erection of plumes varies greatly, but even in low intensity forms the crown plumes are always erected. Both sexes show the Stationary Stretch Display, and it figures prominently in nest relief. Figure 6 in Pemberton (1922) shows a high-intensity form of the Stationary Stretch Display. All birds observed performing Stretch Displays had brilliantly colored soft-parts.

Unlike the Extended Stretch Display of the Green Heron, the Stretch Displays of the Reddish Egret show sexual tendencies. The extreme erection of plumage might be considered a hostile element in this display, but I believe that such movements, particularly the vigorous head-tossing, function to emphasize the movements of the head and neck.

Aerial Stretch Display. This display is performed in full flight. Shortly after launching from its perch, the displaying bird extends its head and neck forward and upward, bill at about a 45 degree angle, and then it head-tosses vigorously. As the displaying bird approaches the object of the display, it raises its head to the vertical and increases the frequency and amplitude of head-tossing. Each toss is accompanied by a soft, not unpleasant "crog" call. Both sexes show the Aerial Stretch Display, and mutual bouts are very common, especially when the pair have been associating for several days.

A possible hostile component, if indeed it is a hostile element, occurs

FIGURE 36. The Stationary Stretch Display of the Reddish Egret.

in the aerial form of the Stretch. Occasionally a bird will bill-snap during flight, and it alternates such snapping with head-tossing and "crog" calling. This might be a low intensity hostile element, but the display is primarily sexual. Alternatively, bill-snapping in the Aerial Stretch Display may have been borrowed from hostile snapping (for example, in Pursuit Flights), and now serves to emphasize a sexual display.

Circle Stretch Display. This is the most bizarre of the sexual displays. I am convinced that this form is performed solely by the males, although no birds were marked. If so, this is the only sexual display of *rufescens* which is performed by one sex only.

The usual sequence of events is as follows: a male and a female are standing quietly some distance apart on an open shoal. Suddenly the male launches into an Aerial Stretch Display, flying directly toward the female. As the male approaches, the female responds with a low intensity Stationary Stretch Display. Immediately upon landing, the male circles the female at a distance of from five to ten yards, but as the intensity of his display rises, he draws closer and closer until he almost touches the female with his extended wings. During his circling the male alternately raises his head to about a 45 degree angle, head-tossing vigorously, then depresses his head until his bill touches the surface of the water. He holds his wings fully extended, alternately extending and closing one wing, usually the one closest to the female. As the male shortens the diameter of his circle, the intensity of his movements increases. Now his head-up motions carry the head and bill over backward, then the downward movements cause his head and bill practically to touch the water. All this time the female has been motionless except for rather low intensity head-tosses. The Circle Stretch Display is usually ended when the male rises to a standing position, very close to the female, and the pair then engage in mutual Stationary Stretch Displays.

I believe that a male performing a Circle Stretch Display is expressing an extremely strong sexual tendency, exceeded only by copulation. The male appears to be ecstatic during such a performance, and the female never lets her attention leave the circling, bowing, head-tossing male. During such a spectacular performance the two birds are easily approached.

Peering Down Display. The causation of this unusual display is obscure. A pair which have been displaying mutual Circle Flights, Stationary and Aerial Stretch Displays, and other sexual activities, will be seen frequently to stand side by side in very shallow water. The soft-parts of both sexes were always brilliantly colored. The motionless pair will then bend over, extend their heads down, bills almost touching the water, and then they will peer into the water, sometimes for a few moments but typically for a minute or two.

The first time I saw this display, I thought that one of the pair, or both simultaneously, had seen some prey close to its feet. However, the highly stereotyped bending, immobility in the bent-over attitude, and the complete absence of any feeding movements tend to eliminate feeding motivation.

Occasionally one of the pair will bend and peer first, but the partner follows suit almost immediately. Because it is associated with sexual displays before and after, I believe that mutual peering down is a sexual display. Furthermore, I saw no indications of hostility before or after this display.

Jumping Over Display. Jumping over has already been described during dancing ground activities, but the present display differs in that both of the pair show it. It appears to express sexual tendencies when shown mutually. Suddenly one of a pair which has been standing quietly in shallow water flies up and over its mate. During the jump-flight, the displaying bird head-tosses vigorously and erects the feathers of the head, neck, pectoral, and scapular regions fully. It may repeat jump-flights as many as ten times in succession. During this display the partner toward which the performance is directed shows high intensity head-tossing in an upright position. In contrast, the stationary bird in dancing ground activities responds by assuming an Upright Display. I believe that the jump-flight during dancing ground activities is of low intensity because it does not involve any feather erections or head-tossing.

After one of the partners has displayed for a few minutes, the other usually jumps over in return. Mutual jumping over may last for ten or fifteen minutes, and it may be followed by a mutual Circle Flight and several mutual Aerial Stretch Displays. I conclude that a strong sexual tendency is being expressed in this display; it is usually preceded and/or followed by other sexual displays. In this mutual display I observed that both birds had brilliantly colored soft-parts.

Circle Flight Display. This display closely resembles the Circle Flight of the Green Heron. The pair is perched quietly in an exposed situation; suddenly one of the partners launches into flight, closely followed by the mate. They fly almost invariably in a circle. One or both birds may show low intensity head-tossing in flight, and several times I heard both birds utter very soft "crog-crog" calls. On their return, the pair may engage in Stationary Stretch Displays and other mutual sexual activities, or they may launch into another Circle Flight Display. Circle Flights are most common during the early stages of pair formation, and I believe that they are expressions of rather low sexual tendencies. No nipping, bill-snapping, or other hostile behavior was noted during this display.

Twig-passing Display. Unlike that of the Green Heron, the Twig-passing Display of *rufescens* is highly ritualized. Either member of the pair may act as "gatherer" or "receiver." Allen (1954–55) describes this display as follows:

> "The reddish egrets, already paired, were getting their nests in order quite rapidly. One member of the pair (usually the female, as determined by subsequent observations) occupied a perch convenient to the nest site while the other (usually the male) wandered about gathering nesting material. When the male returned, with, or quite frequently without, a stick or small branch, he began his recognition and greeting ceremony while still on the wing. This consisted of a series of "back bends," as I called them in my notes. Without missing a wingbeat he raised his head so that the bill pointed upward, neck arched and head bent back, the stiff feathers of the mane only slightly extended. He did this several times so that the effect was one of a series of nods or bows. The female, for her part, raised the mane from crown to breast and pointed her bill straight towards the zenith. As they joined each other on the top of a cactus plant, both birds fluffed out their manes, rubbed their bills together excitedly and uttered low "chicken-like" notes and a flat snapping sound apparently produced by clacking the upper and lower mandibles together. At the same time they kept up a mutual display of short, grave, head nods, one to the other. Generally this was followed by a period of working on the completion of the nest."

Where Allen says "recognition and greeting ceremony," I would substitute "Aerial Stretch Display," and where he uses "a series of nods or bows," I use "head-tossing." In Allen's description the display of the female on the nest would be a Stationary Stretch Display, using my terminology.

In general my observations of the Twig-passing Display of the Reddish Egret agree closely with Allen's account. Observations of the nest-building behavior of three pairs of *rufescens* on Cowpens Key, however, showed that the sexes played equal roles in gathering, receiving, and weaving twigs into the nest. Also, the gatherer began his or her head-tossing while wading in water near the nest. A return to the nest in flight was not invariable. At one nest on Cowpens Key, located only a few feet from the edge of the key and close to the surface of the water, the returning bird almost always returned to the nest walking directly, then awkwardly scrambled up the branches of the nest tree. In every such case the gatherer began its head-tossing immediately after it had seized the twig, and this display continued until the twig had been passed to the mate. Even as the returning bird scrambled and fell among the lower branches of the nest tree, it continued to display. As Allen suspected, the flat snapping sound is made by the rapid movement of the mandibles against each other. Bill-snapping in Pursuit Flights is purely a hostile element, but I believe that bill-snapping in the Twig-passing Display, as in the Aerial Stretch Display, is not a hostile component.

I conclude that the Twig-passing Display of *rufescens* is a variant of the Stretch Display which has been carried over into the nest-building stage of the breeding cycle. *Virescens* and *herodias* also show Stretch Displays during the nest-building stage.

Movements of the head and neck play a prominent role in the vast majority of sexual displays of the Reddish Egret. It is no accident, in my opinion, that the feathers of the head, neck, and pectoral regions of the dark birds are sharply set off from the rest of the body plumage in over-all coloration and plume type. I agree with Lorenz that such feather coloration and structure have evolved to emphasize the exaggerated movements of these regions. Even though the white birds lack this color separation, the bristly plumes and their erection during sexual displays emphasize the upper parts of the body.

EARLY PAIRING BEHAVIOR

Territory. After the dancing ground activities, the males begin to perch for long periods in particular parts of the potential nesting areas. At first these males, always brilliantly colored, threaten and/or attack and pursue other Reddish Egrets which approach their perch. The first response is hostile. The initial area defended is large, but in a matter of days the male's territory has been reduced considerably. Unlike the Green Heron, the Reddish Egret males show Stationary Stretch Displays almost from the beginning of territoriality.

Displaying, threatening males are approached by the females, and the response of the male is at first hostile. Soon, however, the male begins to perform low intensity sexual displays, and then the first mutual sexual displays begin to appear.

The striking feature of the courtship at this early stage is its mobility. Although the male typically returns to his territory after aerial displays, mutual displays may be performed as much as one-quarter of a mile from the male's territory. This mobility may be "permitted" by the open nature of the habitat and by the small breeding population in the Keys, both of which reduce contacts between courting pairs. It would be of great interest to determine the mobility in a crowded population, such as the one on Green Island, Texas.

The first entrance of the female into the nest was the moment of pair formation in the Green Heron. I was unable to determine the exact time of pair formation in the Reddish Egret, but it is my impression that when the pair begin mutual Jumping Over and Peering Down Displays, they are paired. I could not detect hostility between the sexes after this stage. Twig-passing and mutual Stationary Stretch Displays at the nest complete the pairing stages.

Coloration of soft-parts. Table 18 lists the breeding and non-breeding soft-part coloration of *rufescens*. It is my impression that the coloration of the males is more intense than that of the females. The males appeared to acquire their coloration earlier than the females. Allen (1954–55) has published a detailed description of the soft-part colora-

TABLE 18

SOFT-PART COLORATION OF THE REDDISH EGRET.

Area	Coloration
	Non-breeding
bill	distal third black; proximal two-thirds pale flesh-color to pale violet
lores and orbital skin	pale flesh-color to pale violet
legs	blue
feet	blackish
iris	white to pale straw
	Breeding
bill	distal third lustrous jet-black; proximal two-thirds brilliant pink
lores and orbital skin	brilliant violet-blue
legs	brilliant blue (somewhat more greenish-blue in some white phase birds)
feet	blackish
iris	white to pale straw

tion of the Reddish Egret, and I agree with his statement, "the soft parts are striking in color and apparently uniform in all plumage phases." In the Florida Bay region, however, several white phase birds in breeding coloration had legs somewhat more greenish-blue than bright blue.

SNOWY EGRET

Study areas, methods of observation, and other details were the same as with the other species. Occupation of the breeding areas was not studied in detail, but in my experience *thula* resembles *albus* in that the birds moved into the nesting areas over a long period extending weeks after arrival on the breeding grounds.

DISTRIBUTION AND DESCRIPTION

As a breeding bird the Snowy Egret, *Leucophoyx thula,* ranges from California, Idaho, and Colorado in western United States east to Long Island, New York, and Massachusetts, south to Mexico, the Gulf Coast of the United States, and through Central America to Chile and Argentina in South America. In recent years this egret has been steadily expanding its breeding range northeastward in the United States. At the periphery of its breeding range the Snowy Egret may nest singly, but large colonies are typical.

The Snowy Egret may be recognized by behavior, size, and pure white plumage. It is one of the smaller North American herons, about twenty-five inches in length, with a wingspread of about three feet. The plumage is immaculately white, the bill is entirely black, as are the legs, while the feet are a bright yellow, except during the breeding season. In the nuptial plumage the Snowy Egret has a crest of short, aigrette-like plumes; long aigrette scapular plumes recurved at the tips; and it has similar plumes, but not recurved, in the pectoral region of the ventral tract. These unusual plumes figure prominently in many hostile and sexual displays.

MAINTENANCE ACTIVITIES

Locomotion. In flight *thula* closely resembles *virescens* in the typical upward and forward extension of the head and neck just before landing, and in the almost invariable crest-raising just before alighting. Snowy Egrets frequently erect their crests when taking off, as do Green Herons.

The flight is rapid and direct, and *thula* appears to beat its wings more rapidly than do the other North American herons. The Snowy Egret resembles *rufescens, herodias,* and *albus* in that it walks rather slowly, except when feeding, in an upright position, unlike *virescens* which walks from a low crouch. In its other locomotory activities *thula* does not differ from the larger herons previously described.

Feeding behavior. The strenuous activity of the Snowy Egret when feeding has been observed by a host of authors (for example, Audubon, in Bent, 1926; N. B. Moore, in Baird *et al., op. cit.*; Forbush, 1929). Even though it is typically very active, I have seen *thula* use the basic heron feeding types Stand and Wait and Wade or Walk Slowly on many occasions. Sprunt and Chamberlain (1949), speaking of the Snowy Egret, state, "It is no still-hunting bird." I have observed individual Snowy Egrets perched for long periods on logs and mangrove roots, and I have watched them feed frequently by wading very slowly. Their more active hunting techniques, however, are usually seen.

Wing-flicking is a very common feeding technique, and a Snowy Egret when feeding in this manner closely resembles *rufescens*. That is, the bird moves forward slowly or at a rapid pace, then periodically extends and retracts its wings in a rapid flicking motion. The wing-flicking frequently shifts into Open Wing feeding, where the bird runs forward, or dashes wildly about, both wings widespread, and then stabs to the right and left. When feeding in this manner, *thula* closely resembles *rufescens* and *tricolor*.

I have never seen a Snowy Egret use Canopy Feeding, but it will use a stationary open wing stance where the bird comes to a halt from a rapid run, fans one wing, and then stabs at any prey disturbed.

One of the most characteristic feeding methods used by *thula* is foot-stirring. In Florida, "This heron (that is, the Snowy Egret), more than any other kind, is what Mr. N. B. Moore designates as a scraper, or raker, because it uses its legs and claws to start from their hiding-places such animals as it wishes to seize for food—namely, crawfish, tadpoles, suckers, aquatic insects, etc. In this movement it far surpasses all other species, and manages its legs with greater adroitness and rapidity" (Baird *et al.*). My observations of the foot-stirring of *thula, rufescens,* and *tricolor* support Moore's statement (Meyerriecks, 1959).

Numerous observations of the foot-stirring of *thula* have been recorded (for example, Walsh, 1929; Jones, 1936; McIlhenny, 1936; Hickey, 1937; Haverschmidt, 1948a; Rand, 1956), and most of the descriptions agree in general. The feeding bird extends one leg, then rapidly vibrates it so as to impart a raking, scraping, shuffling, or stirring motion to the foot, and then snaps at any prey disturbed.

The bright yellow toes of *thula* are sharply set off from the black legs, and this distinctive pattern may have evolved in conjunction with foot-stirring. It is of interest to note that the Little Egret, *Egretta garzetta* (Linnaeus), of the Old World, also has dark legs and bright yellow feet, and that this species also engages in foot-stirring (White, 1947; Hobbs, 1957). However, the feet of *thula* turn a brilliant coral-orange and those of *garzetta* turn a brilliant crimson-pink during the early part of the breeding season, and the feet of *thula* are prominently displayed during courtship, so the primary function of the distinctively colored feet is not entirely clear. Other details of foot-stirring, hovering-stirring, and foot-paddling in the Snowy Egret are to be found in Meyerriecks (1959).

Rice (1954, 1956) has described the close association of Snowy

Egrets with cattle as they fed on grasshoppers and other insects disturbed by the grazing animals. This behavior is similar to the association of Cattle Egrets, *Bubulcus ibis* Linnaeus, and Little Egrets with cattle and other grazers. Stoddard (in Howell, 1932) observed a flock of about forty Snowy Egrets following a drove of pigs on a Florida prairie and picking up grasshoppers disturbed by the pigs.

Care of the body surface. The Snowy Egret agrees closely with the Reddish Egret in the frequency of underwing preening. This behavior may well be correlated with the use of the wings in a variety of activities. Through the courtesy of Samuel A. Grimes I am able to reproduce his beautiful photograph of underwing preening by the Snowy Egret (Plate 15). The Snowy Egret does not differ from the species already described in any significant details in other maintenance activities.

Alarm, "freezing," and escape responses. In its alarm and escape responses the Snowy Egret resembles the Green Heron more than any other North American heron. Like the Green Heron, *thula* shows moderate to extreme crest-raising in a variety of alarm situations. The Snowy Egret also agrees with *virescens* in that crest-raising is most frequently seen from one of the extended alert-postures, that is, with head extended fully upwards. Where *virescens* shows rapid crest-erection and depression from this attitude, however, *thula* typically maintains the crest in an erect position, only slowly depressing the crest as the source of the alarm (a human, for example) departs.

When flushed suddenly, *thula* will erect its crest to an extreme degree and utter several loud, rather high-pitched "aarg-aarg" call notes. Such calls typically alert all other Snowy Egrets within hearing range (a social warning signal).

Alert postures in *thula* agree closely with those of *virescens* and other herons in that the thinness and extension of the neck are the keys to the strength of the escape tendency.

With respect to one other escape response, the Snowy Egret tends to behave like the Green Heron. Whenever a Snowy Egret was suddenly surprised by the appearance of a Marsh Hawk, the egret adopted a low intensity Bittern Stance. The bird momentarily froze with its neck extended fully, but unlike *virescens*, it held the bill up at about a 45 degree angle. Green Herons hold the Bittern Stance for a long period, but the egrets adopted this pose only for a few moments and usually flushed when the hawk approached closely. To be anthropomorphic for a moment, it appeared as though the egrets had momentarily "forgotten" that they were white and most conspicuous.

HOSTILE BEHAVIOR

Non-aerial, Non-vocal Displays

Crest-raising. The crest is erected in many displays, both hostile and sexual. The non-aerial displays in which the crest is erected are the Upright (erection moderate to full), Aggressive Upright (always full erection), Forward (extreme erection), and the Snap Display (full erection). During fights, which are common, the crest is typically erected to an extreme. The degree to which the crest is erected in any hostile display is a good indication of the strength of the attack tendency, as is the case with *rufescens.* I believe that the crest-raising shown in the various alert postures of the Snowy Egret is an expression of conflict (see discussion of the causation of crest-raising in the Green Heron, p. 26).

The Green Heron and the Reddish Egret resemble each other in that the crest is the most mobile feather unit. However, the Snowy Egret differs in that the crest, pectoral plumes, and the scapular plumes all appear to be linked in their mobility. That is, whenever one of these feather units is erected, then the other two are erected also, particularly during hostile displays, and in sexual displays also. These facts suggest that erection of these modified plumes in hostile displays increases the apparent size of the displaying bird, while in sexual displays the feather erections emphasize certain movements of the body.

Tail-flipping. From my limited observations, tail-flipping does not appear to be a *characteristic* component of hostile displays. I have seen Snowy Egrets flip their tails in the Upright, Aggressive Upright, Forward, and Snap Displays, and I saw tail-flipping during some extreme alert postures, but it did not appear to be a characteristic element.

Upright Display. A Snowy Egret in this display shows the following movements:

1. extends the head and neck upward to their fullest extent, with the neck held vertical
2. holds the bill typically horizontal, but may incline it slightly downward
3. erects the crest partly or fully, and erects the pectoral and scapular plumes moderately
4. holds its wings tightly closed, and may flip the tail infrequently
5. gives no sound, or may utter several rather harsh, low-pitched "aarg" or "aarh" calls.

The Upright Display of *thula* resembles the Upright of *rufescens* except for the vertical position of the head and neck and the erection of the pectoral and scapular plumes. I conclude that the Uprights of these two species agree closely in their causation; that is, the escape tendency, although actually low, is relatively stronger than the attack tendency. A slight increase in the escape tendency leads to orientation away from the opponent, or unmistakable intention movements of flight (sleeking of plumage, retraction of head and neck, and leg-bending). In contrast, a slight increase in the attack tendency leads to the next higher expression form, the Aggressive Upright Display.

The Upright is a common sight, especially during the very early stages of pair formation and territoriality. Occasionally the threatening birds will move very slowly toward each other, both birds in the Upright, but typically one or both birds will assume the Aggressive Upright as they move closer to one another.

Aggressive Upright Display. As a bird assumes the Aggressive, it shows the following components:

1. extends head and neck upward to their fullest, and inclines them slightly forward from the vertical
2. inclines the bill downward at about a 45 degree angle
3. erects the crest fully, and erects the pectoral and scapular plumes either moderately or fully
4. bulges its eyes from the head
5. holds its wings either tightly closed or partly extended
6. typically utters several harsh, rasping "aarg" or "aarh" calls
7. remains stationary or moves very stiffly toward the opponent.

The Aggressive Uprights of *thula* and *rufescens* are similar except that forward inclination of the neck is more extreme in *thula*.

Forward Display. This is the most intense of the hostile displays, and when assuming it the bird shows the following components:

1. retracts its head and neck slightly, so that the occipital region is directly over the upper part of the back, neck in an "S"-shape
2. inclines its body forward as it bends slightly at the "heel" joints
3. erects its crest, pectoral and scapular plumes to an extreme, the tips of the scapulars almost touching the fanned plumes on the back of the head
4. inclines its bill downward at about 45 degrees
5. bulges its eyes from the head
6. partly extends its wings and frequently "fans" them slightly
7. gives harsh calls, typically rasping "aargs," or an even harsher "raah" call
8. faces the opponent, either stationary or while moving stiffly toward the opponent
9. may show color changes in the lores and orbital skin in a matter of seconds (Figure 37).

A Snowy Egret in the Forward Display is "all feathers." The apparent body size is very much increased by the extreme erection of all the plumage, especially the highly modified plumes. The retraction of the head may express an increasing escape tendency, but I believe that this makes the bird ready for a strike at the opponent.

The change in color of the lores and orbital skin is unusual. The Snowy Egret is the only heron I have seen undergo such rapid changes in soft-part coloration. The lores and associated areas may change in a few seconds from yellow to orange to cherry-red. I believe that these rapid changes are correlated with the increasing attack tendency because the cherry-red rapidly shifts to orange or yellow once the bird has ceased the hostile display. In addition, I have seen such rapid color changes only during hostile clashes, never during sexual displays. Lowe observed rapid color changes in the bill of *A. cinerea* on its dancing grounds, and this same author cites George Edward's observa-

FIGURE 37. Snowy Egret performing the Forward Display.

tion of rapid color changes in a Little Bittern, *Ixobrychus minutus* (Linnaeus), during a nest relief ceremony.

Fighting. I have observed more fights in *thula* than in any other North American heron, even though I have spent far less time watching *thula* than *virescens.* Typically, two fighting Snowy Egrets fly headlong toward one another from the Forward Display, collide in the air, and then tumble to the ground. Characteristically, they attempt to strike their opponents from above, aiming their vicious jabs at the head. They flap wildly trying to buffet each other. Occasionally one of the fighters jumps into the air and tries to land on the other's back, and the jabbing, whirling, flapping opponents may range over a hundred yards or more in their struggles. During a fight, as Storer's (1948:89) photographs show, the birds may actually fly backwards. Fights rarely lead to physical damage except for the loss of one or more plumes.

Both the Reddish Egret and the Snowy Egret fight from an upright position, strike from above, fight furiously and over a wide area, and enter the engagement with a headlong dash.

Snap Display. The Snowy Egret, like the Green Heron, shows "*cinerea*" type Snap Displays (no bobbing components) as well as variants in which the bird bobs rapidly up and down. The Snap Display of *thula* which lacks the bobbing motions has the following components:

1. the bird extends its head and neck fully forward and downward so that the head and bill are on a level with or slightly below the feet
2. raises the crest fully, and the feathers of the neck partly or fully
3. makes a loud snapping sound by clicking the mandibles together as it extends its neck
4. bulges the eyes from its head
5. may flip its tail, but not characteristically
6. holds the Snap Display position for a few moments, then rises to a normal perched position.

The bobbing variant is as just described except that the displaying bird, with its neck fully extended, bobs rapidly up and down and gives the neck a peculiar bowed appearance during the bobbing movements. The Snap Displays of *thula* most closely resemble those of *virescens,* but the Snap of *thula* appears to be much more of a hostile display because it appears only early in the breeding season, because the displaying bird alternates Snaps wth much more intense hostile behavior (Aggressive Uprights and Forwards), and because conspecific individuals are attacked if they approach the displaying bird. Furthermore, Green Herons tend to confine bouts of Snapping to their own territories, but Snowy Egrets may Snap on or off their territories.

In short, the form of the Snap Displays resembles those of *virescens,* but causation and orientation resemble those of *rufescens.*

Non-aerial, Vocal Displays

The characteristic advertising call is a nasal, fairly high-pitched "aarh," uttered repeatedly from a prominent spot. One variant I heard was a rather high-pitched "arg," very similar to the alarm call. Another advertising call, heard most frequently before the performance of sexual displays, sounded like "arg-obble." The "obble" sounded like a low intensity form of a call that characteristically accompanies the Stretch Display.

Although I collected and sexed no Snowy Egrets, it is my impression that both sexes utter the "aarh" and "arg" advertising calls. I believe, however, that the "arg-obble" variant is restricted to the males.

The Snowy Egret is highly social throughout the year, and extreme sociality carries over into the early pairing period. Groups of from five to ten egrets engage in advertising bouts, all of them being perched in one small area, whereas Green Herons engage in "skowing" bouts with one (rarely more) neighboring territorial male. Unlike the Green Heron, the Snowy Egret appears to have an ill-defined territory, at least during the first few days. Typically, one male begins to advertise from a conspicuous song post, and then he is joined by several males and females. A bout of calling back and forth ensues, and the little party of advertisers moves about, not settling on one particular spot for very long. Again unlike *virescens,* advertising in *thula* is highly social.

Aerial, Non-vocal Displays

Flying Around. Flying Around is strikingly similar to that of *virescens,* except again that *thula* displays are highly social. Single birds show this display, but it is much more common to see small groups of from five to ten egrets Flying Around. Most of the birds engaging in this display show signs of changes in soft-part color, the feet especially being a brilliant coral-orange. A bout of Flying Around may take an individual or the group over hundreds of yards, and unlike the Green Heron, which usually returns to the starting point, these little flocks may settle far from their starting point. Later in the breeding cycle, Flying Around bouts become restricted to smaller and smaller areas, and eventually the flights end at the take-off spot.

The function of Flying Around in *thula* appears to be the same as in *virescens* and *rufescens.*

Pursuit Flight. Social Pursuit Flights are much more common than flights which involve only two birds. During such a social flight the pursuer may chase first one bird then another, only to be pursued in turn by some other individual. Pursuit Flights involving two birds, however, are not unusual. Most Pursuit Flights of two birds result from the too close approach by one bird to an advertising individual; the latter first threatens the intruder and then pursues it. In such cases the pursuer typically returns to a place close to its former song post.

During the social Pursuit Flights crest-raising, "arg"-calling, and nipping in flight are very common. The social nature of such flights makes their causation difficult to analyze, but the occurrence of crest-raising, calling, and nipping indicate that the flights are primarily hostile.

Supplanting Attacks. This, of all the hostile displays, most closely resembles the same display in *virescens.* The upward and forward extension of the neck of the supplanting bird, and the occurrence of crest-raising in both individuals agree closely in the two species. In addition, I saw Snowy Egrets make incipient copulatory attempts during Supplanting Attacks, just as male Green Herons do on occasion. In both species Supplanting Attacks are most frequent early in the breeding season, their intensity and frequency falling off rather sharply as the breeding season advances.

Aerial, Vocal Displays

Although I heard individual Snowy Egrets call "arg" or "aarh" during Pursuit Flights or Supplanting Attacks, I saw no hostile, aerial, vocal displays which appeared to be ritualized.

SEXUAL BEHAVIOR

With respect to three displays, the Stationary Stretch Display, the Aerial Stretch Display, and the Jumping Over Display, the Snowy Egret resembles the Reddish Egret more than any other North American heron with which I am familiar.

Stretch Displays. The Snowy Egret shows two distinct Stretch Displays, a stationary one and an aerial version, and there are several low intensity forms of both types.

Stationary Stretch Display. From a very conspicuous perch the displaying bird performs the following movements:

1. raises the head in one smooth, rather rapid motion, then brings the head back and downward so that the occipital region is directly over the lower part of the back
2. points the bill straight upwards
3. erects the feathers of the crown moderately, erects the pectoral plumes fully and fans them laterally, and erects the scapular plumes to an extreme so that their recurved tips touch the erected plumes of the crest
4. bulges its eyes from the head
5. bends the legs at the "heel" joints, then begins to pump its head vigorously up and down, the occipital region touching the lower part of the back on the downstroke
6. as the bird pumps its head, it utters a series of gurgling "a-wah-wah-wah," "a-wah-wah-wah" calls (Figure 38).

In low intensity forms of the Stationary Stretch Display the bird may utter only one or two "a-wah-wah-wah" calls or none at all, and

FIGURE 38. Male Snowy Egret performing a Stationary Stretch Display.

it may reduce the number of pumps of the head to one or two. In full intensity displays the bird may pump ten or more times and utter a long series of calls.

I have seen the Stationary Stretch Display performed only by what I took to be males. The extreme erection of the scapular plumes certainly emphasizes the pumping movements of the bird's head, and the brilliant cherry-red lores and orbital skin draw the observer's eyes toward the rapidly moving head. In addition, the laterally-fanned pectoral plumes and the brilliant coral-orange feet, sharply marked off from the lustrous black legs, emphasize the exaggerated movements.

The "a-wah-wah-wah" call which accompanies this display is a rather low-pitched, not unpleasant gurgling call which may also be uttered by the bird just before performing the Stationary Stretch Display. In fact, the bird may give this call, or a less intense variant, rather frequently days before it first performs the Stationary Stretch Display.

One of the most interesting aspects of this display is its amazing attraction function. As soon as a male assumes the Stretch Display position, Snowy Egrets within sight and hearing fly to perches near the performing bird, and all of these "spectators" watch the performing bird intently. I have seen one displaying male ringed by as many as ten other "fascinated" Snowy Egrets. These gatherings also indicate the degree of sociality found in this species.

Aerial Stretch Display. From a conspicuous perch, the displaying bird launches into flight as follows:

1. the bird stands erect, neck partly extended forward at about a 45 degree angle; crest and scapular plumes are slightly erected
2. it bends forward from this position so that the body is almost horizontal, neck extended forward fully, crest moderately erect, scapular plumes fully erect, and the wings slightly drooped
3. droops the wings strongly, and then raises its head upward and begins to utter the "a-wah-wah-wah" call
4. from this position the bird launches into flight, sways the head and neck from side to side several times, then brings the head and neck down and to the rear into the Stationary Stretch position except that the bill is pointing forward at about a 45 degree angle
5. the bird then flies slowly in a large circle, continuously calling the "a-wah-wah-wah" call with its head and neck back, all of its plumes, but especially the scapulars, streaming in the wind
6. about three-fourths of the way around the circle, the bird brings its head and neck forward but still partly extended, lowers its plumes, and stops calling
7. just before landing, the bird makes a sudden short flight upwards, dangles its legs, elevates its crest and scapulars, then "pounces" down to its perch (Figure 39).

Everything about this spectacular performance is highly exaggerated, and all of the plumes, soft-part coloration, and extremely bizarre movements, together with the continuous calling, render this a unique display. I believe that only males show this aerial performance, but

FIGURE 39. Snowy Egret performing an Aerial Stretch Display.

unlike the stationary version, this display is performed before a single female. She follows every movement of the performing male intently all through his flight until he lands near her (following the short up-down flight).

I have seen the Aerial Stretch Display only after a male and female had been engaged in a variety of activities together for several days, and it is my impression that the aerial form is an expression of much stronger sexual tendencies than the stationary type. During the aerial performance the displaying male appears to be ecstatic, and the female is so intent on watching the male's activities that she may be approached rather closely.

Circle Flight Display. The Snowy Egret may engage in individual or mutual Circle Flight Displays, and the performances are very similar to those of the Green Heron.

When performing singly, the displaying bird launches into a typical flight, flies in a circle around the mate, and then lands near the starting point. Occasionally the bird may utter one or two "a-wah-wah-wah" calls during flight, but this is not typical. Several times I have seen a bird performing this display raise its head slightly in what appeared to be low intensity form of the Aerial Stretch Display, but these movements are very slight.

In mutual Circle Flights one of the partners launches into flight and is typically followed at once by its mate. The two birds fly in a circle around the take-off spot and then land near this point. Sometimes the last bird to fly overtakes its partner in flight, but I have never seen any signs of hostility on such occasions. Individual Circle Flights are much more common than mutual ones early in the breeding cycle, but once a male and a female have been associating for several days, mutual Circle Flights are the rule. Such flights are sharply reduced in frequency following the first performance of an Aerial Stretch Display.

Tumbling Flight Display. One very outstanding display, which appears to be a variant of the Circle Flight Display, involves some spectacular aerial maneuvers. The performing bird launches into a typical Circle Flight, but instead of "once around, then land," the bird continues to circle its mate, rising higher with each turn. When it has reached a point about fifty to a hundred yards (very variable) above the ground, the bird suddenly tumbles over and over as it hurtles downward, all of its plumage streaming in the wind. The displaying bird may actually turn completely over three or four times during its fall. After tumbling, the bird rights itself, then lands in a standard manner near its partner. I have seen such tumbling flights outside the breeding season when Snowy Egrets were approaching a roosting site. This species characteristically approaches a roost flying high over it and then, twisting and whirling, plunges downward to the tops of the trees, "braking" to a sudden stop.

Jumping Over Display. This display resembles closely the Jumping Over Display of the Reddish Egret, except that the Snowy Egret does not show any head-tossing. Typically, a pair of Snowy Egrets engage in a mutual Circle Flight Display, but instead of returning to the take-off spot they land in the shallow water (Florida Bay) or on the marshes (Rulers Bar Hassock) near the nesting sites. Both stand quietly for a few moments, then suddenly one makes a short jump-flight over the other, lands, then repeats the jump-flight. During the jump-flight the displaying bird erects its crest, pectoral, and scapular plumes fully, and these feathers stream and toss about as the bird moves up and down over its partner. Sometimes a few jumps by one of the pair ends the display, but I have seen the partner display in return on several occasions. The behavior of the pair after a bout of jumping is very variable; sometimes they start to feed, or they may return to the breeding sites, or they may stand quietly for several minutes before flying away. The fact that this display is shown by a pair which has been engaging in mutual activities for several days, plus the fact that no hostile behavior was noted, leads me to conclude that it is a sexual display.

Sherwood (1957) makes the only previous reference to what appears to be this display: "One day in January [in southern Florida] three pairs of Snowy Egrets put on a mating display in the now shallow water, jumping over one another with plumes flying. . ."; however, no further details are given. It would be most interesting to discover whether or not Snowy Egrets engage in social Jumping Over Displays, as Sherwood's observation suggests.

EARLY PAIRING BEHAVIOR

Territory. The establishment and maintenance of territory by the Snowy Egret resembles such behavior in the Reddish Egret, except that territorial advertising is much more conspicuous. With respect to the gradual occupation of the breeding areas, *thula* resembles *albus* closely.

The first indications of breeding behavior are the return of certain individuals from the feeding areas early in the day to the potential breeding sites. These birds perch quietly in conspicuous spots. Most have bright orange feet and yellow lores. An occasional bout of Flying Around alternates with prolonged preening, but these birds are very alert to the appearance of other Snowy Egrets. Should one land in the breeding areas, the perched birds typically Supplant the newcomers, and Pursuit Flights are common. Rarely, one of these birds engages in a low-intensity Snap Display. Half-hearted twig-grasping also occurs, but the twigs are merely grasped and then released at once.

As soon as several individuals spend more of the day in the breeding areas, the first advertising calls are heard. Bouts of "aarh" or "arg"-calling become common, and the first hostile clashes occur. At times one of the calling birds shows a low intensity Stationary Stretch Display, thereby attracting a small party to it. The usual response of the displaying bird is to threaten the "audience," whereupon the latter retreats until the attacker again displays a Stationary Stretch. This gathering and retreating is especially conspicuous during the early phases of the breeding cycle.

Soon after the appearance of the first Stationary Stretch, individual Circle Flights become common, and occasionally one of the flying birds will show a Tumbling Flight Display. As more and more egrets move into the breeding areas, threats and fights increase in frequency, and the original territories of the first arrivals become restricted. On Rulers Bar Hassock an old nest was not necessary for the establishment of territory, in sharp contrast to the Green Heron.

Once the male has selected a spot, his Stationary Stretch Displays rise in frequency and intensity. Now, one or two birds of the small groups that gather retreat very slowly when the male threatens, and I took these individuals to be females. The hostile displays of the male decrease to the point where these females are no longer threatened, and then the male performs his first Aerial Stretch Display. At this point the females begin to confine their attentions to one displaying male rather than flying from one gathering to the next.

Soon the pair begin to perform mutual Circle Flights, and the male shows more and more Aerial Stretch Displays. Advertising calls fall off abruptly and aerial displays, such as the Tumbling Flight Display, rise sharply in frequency and in intensity. I believe that pair formation is effected when the pair show mutual Jumping Over Displays. After such performances the pair may return to the nesting site and engage in nest-building activities. I made no observations of the pair after this stage.

Coloration of soft-parts. Table 19 lists the breeding and non-breeding coloration of the soft-parts of the Snowy Egret. No sexual differences are listed, but I believe that the males show more intense coloration and acquire their color changes earlier than the females; this is especially true for the changes in the color of the feet. Huxley (1922), as far as I have been able to determine, was the first to note the change in lore color from yellow to red during the breeding season. Cruickshank (1948) has recorded the cherry-red coloration of the lores of *thula* early in the breeding season, and several authors have commented on the orange appearance of the feet of breeding birds, but no detailed descriptions of such changes have been published previously.

I have already described the rapid change in color of the lores and

TABLE 19

SOFT-PART COLORATION OF THE SNOWY EGRET.

Area	Coloration
	Non-breeding
bill	black, basal third somewhat yellowish
lores and orbital skin	bright yellow
legs	black
feet	bright yellow
iris	yellow
	Breeding
bill	shiny black throughout
lores and orbital skin	cherry-red
legs	shiny black
feet	brilliant coral-orange
iris	yellow

orbital skin of this species in the Forward Display. Such rapid changes seem, however, to be restricted to the early phases of the breeding cycle, because birds in Stationary and Aerial Stretch Displays normally have brilliant coloration.

When Snowy Egrets first occupy the breeding areas, the lores are of the same bright yellow as they are outside the breeding season, but in a matter of days, the lores have become a bright orange. Later, the lores become a bright cherry-red, and this red color lasts until the clutch of eggs is completed, when it fades. The fading may progress through dull red to orange, then to bright yellow, or from bright red to yellow directly.

The feet also change, from bright yellow to brilliant coral-orange; then there is a gradual fading back to yellow once the eggs have been laid. This shift in color of the feet is an excellent indicator of the onset of breeding behavior. Many of the birds that begin advertising early in the breeding season have bright orange feet but yellow lores.

Part III

SOME EVOLUTIONARY TRENDS IN HERON BEHAVIOR

INTRODUCTION

It is difficult to discuss with any degree of finality the origin and evolution of heron behavior patterns because only a few of the more than sixty species of the family Ardeidae have been studied in detail. For example, the tiger bitterns of Mexico and Central and South America are considered to be the most primitive members of the family. These species would appear to be rather unspeciailzed in their displays and general behavior. The lack of well developed plumes in tiger bitterns tends to support the view that the displays associated with such plumes would be simple in form. In brief, the tiger bitterns would provide significant information as to the probable origin of many of the displays of the more advanced forms. However, the breeding behavior, and indeed the general biology, of tiger bitterns is poorly known. Hence, the student interested in evolutionary origins and trends must use behavioral data gathered from studies on the more advanced species.

I shall discuss in this section some evolutionary modifications based primarily on the behavior of the North American species treated in Parts I and II. Wherever appropriate I shall cite examples of key species whose study would greatly assist in the clarification of many of the points under discussion.

SOCIALITY

As stated in the Introduction, the family Ardeidae is particularly suitable for a comparative ethological study, and numerous examples were given in support of this contention. One was the extreme degree of diversity shown with respect to social behavior. With *A. herodias* and *A. occidentalis* considered to be conspecific, twelve species of herons breed in North America north of Mexico. They exhibit in striking fashion many degrees of sociality. I recognize five social categories with respect to the North American herons (Table 20).

I do not intend to imply that this is the only possible classification of social behavior among herons, or that the species listed are rigidly delimited by one category. For example, solitary Snowy and Cattle Egrets, and Little Blue Herons are commonly observed, particularly at the periphery of their ranges, but these species are much more typically seen in small to large flocks. Furthermore, nesting pairs of such highly social species as the Louisiana Heron (Meyerriecks, 1957a), are com-

TABLE 20

DEGREE OF SOCIALITY OF NORTH AMERICAN HERONS. *Ardea cinerea* IS INCLUDED FOR COMPARATIVE PURPOSES.

Degree	Species
Solitary throughout the year, including solitary breeding	*Botaurus lentiginosus*
Solitary outside breeding season, solitary to semisocial breeding	*Butorides virescens*, *Ixobrychus exilis*
Solitary or semisocial outside breeding season, highly social breeding	*Ardea herodias*, *A. cinerea*, *Casmerodius albus*, *Dichromanassa rufescens*
Mainly social throughout year	*Florida caerulea*, *Hydranassa tricolor*
Highly social throughout year	*Leucophoyx thula*, *Bubulcus ibis*, *Nycticorax nycticorax*, *Nyctanassa violacea*

monly seen singly or in small numbers at the periphery of the breeding range. The true social status of these pioneers, however, is revealed by an examination of the behavior of such a species toward the center of its range.

The probable course of social evolution in herons has been from the primitive solitary status exemplified by *Botaurus*, through a transitory semisocial phase (*Butorides* and *Ixobrychus*), to extreme, year around sociality, represented by *Leucophoyx*, *Bubulcus*, and *Nycticorax*. The Night Herons (*Nycticorax* and *Nyctanassa*) must have evolved their present highly social structure independently of such species as *thula* and *rufescens*, because they differ in numerous other respects, both behavioral and structural.

The change from solitary to semisocial is shown by the American and Least Bitterns, and the Green Heron. The American Bittern is typically solitary throughout the year. On migration or in suitable nesting areas, however, this species may assemble in small groups. Green Herons and Least Bitterns are typically solitary outside the breeding season, and in many areas *virescens* and *exilis* are known only as solitary nesters. On migration, however, small to large flocks are not uncommon, and in suitable areas many pairs (hundreds in *virescens*) may nest in close association.

Those species which are typically solitary or semisocial outside the breeding season provide examples of the next stage in the evolution of a highly social structure. *A. herodias* and *C. albus* are commonly seen feeding singly or in small groups, but on migration, when flying to and from roosting areas, and during the breeding season, both species are highly social. It is but one further step to the full development of a species that is highly social in all of its activities (feeding, migration, roosting, nesting, social displays). For the North American herons this high degree of sociality is best shown by *thula, ibis,* and *nycticorax.*

POLYMORPHISM

Polymorphism, as color phases, is widespread in the family Ardeidae. Of the twelve breeding herons of North America, four have distinct color phases (*herodias, rufescens, caerulea, exilis*). The polymorphism of *caerulea* is developmental, immatures being white while the full adults are dark blue; during the shift from white to dark the changing birds are strikingly pied. *Exilis* has a very rare, melanistic phase, formerly known as "Cory's Least Bittern" (see Carpenter, 1948, for details). The polymorphism of *rufescens* is complex, this species having a minimum of three phases: dark, intermediate, and white (see the treatment of *rufescens* in Part II). Quantitative analysis of the large numbers of *rufescens* breeding on Green Island, Texas, offers an opportunity to study the inheritance of color phases. Such inheritance is not known for any heron species, although Mayr and Amadon (1941) and Mayr (1956) have suggested several genetic models.

The available evidence indicates that color phases do not play a role in choice of mates (Meyerriecks, 1957b, for *herodias*); the functions of such color phases remain obscure. Some species (*albus* and *thula,* for example) have only the white phase. Palmer (1909) claimed that the dark-bodied herons obtained their aquatic prey "almost entirely where the fringe of vegetation, sedges, bushes, or more distant tree tops, make a background and prevent the shape of the birds from affecting the skyline as seen by the prey." In contrast, according to Palmer, the white-bodied species tend to feed in the open, away from the fringing vegetation. I cannot agree with the generalization "dark heron-shore line feeder, white heron-open water feeder," because I have seen on countless occasions the reverse situations, that is, dark-colored herons feeding on fully exposed reefs and shoals, and pure white herons feeding for long periods among mangrove roots and similar vegetation. Palmer believed that the white plumage of the open water feeders harmonized with the sky (as seen by the prey). If this were significant, one would expect that in the smaller species, a two-color pattern would have evolved, white below (to "fool the prey"), and dark above (to protect the heron from aerial predators). In North America *tricolor* is an excellent example of such a species: white below, blue above, and a diurnal feeder. However, *tricolor* may feed among the tangled vegetation lining the shore or out on the most exposed flats. Many of the other small, diurnal feeders are all white (*thula* and *garzetta*), while other small species are completely dark (adult *caerulea*). If white plumage did evolve as a cryptic adaptation, then *caerulea,* white as an immature, dark as an adult, would have to reorganize its entire feeding behavior when its plumage changed. Of course, this is not the case.

Whiteness or near whiteness is not a unique attribute of egrets, because this characteristic is shared by a number of birds (gulls, for example). The coloration of gulls—cryptic or conspicuous?—has been the subject of a recent controversy. Craik (1944) believes that the white color of many gulls is cryptic, while Darwin (1890) and Armstrong (1946), among others, believe that such coloration is conspicuous. For a more detailed discussion of this subject, see Tinbergen (1953). I agree with Tinbergen that only a thorough experimental analysis will resolve the situation.

Finn (1919) linked whiteness in herons with a tropical distribution and conspicuousness, but he gives no details. Several all-white species, such as *albus* and *thula,* range far north of the tropics, but it is true that the white herons tend to be much more common in the warmer regions of the world. Wells, Huxley, and Wells (1929) link whiteness with colonial nesting and relative invulnerability to predation, the advantage being that the white birds are much more conspicuous to conspecific individuals. Koenig (1952) is very definite about the significance of whiteness in *albus.* He states, "Bedeutung des weissen Gefieders zeigt sich deutlich, wenn andere Reiher mittags den Schatten aufsuchen, während die Silberreiher [*albus*] unbekümmert stehenbleiben." However, I have seen white *herodias, albus,* and *thula,* together with immature *caerulea,* retire to the shade at high noon on bright, hot days, while the dark-bodied species, such as adult *caerulea* and *virescens,* remained in the blazing sun for hours. I made observations such as these so many times that I feel that the primary function of whiteness in herons is not correlated with temperature.

The matter of conspicuousness may also be considered in relation to warning colors and edibility, but Cott (1947) has clearly shown that the Cattle Egret is a striking expection to his experimentally based generalization that conspicuousness implies inedibility, while a cryptic coloration indicates edibility. Cott rated his experimental birds on visibility, edibility, and vulnerability to attacks by predators. *Ibis* rated the highest (10) in visibility, 19 in edibility (20, most edible), and 12 in vulnerability (10, relatively immune, 21, highly vulnerable). Hence, whiteness in herons is not correlated with inedibility. Because most herons are relatively immune to predation, at least as adults, it does not matter whether or not their flesh is edible or otherwise.

In a recent description of a new race of *Butorides striatus* from the Maldive Archipelago, Phillips and Sims (1958) express the belief that the increase in white in the Maldive *striatus* is probably correlated with the exposed coral reef habitat of these birds, and these authors assign a cryptic function to such partial albinism. The Striated Herons of western Australia, however, are reddish in areas of reddish soil; hence, the partial albinism of the Maldive birds (which is common in many groups) is not related to the general problem of whiteness in other herons and egrets.

Most of the all-white species of herons, or the white individuals of those species that show color phases, are primarily social throughout the year. Even those white species that tend to be solitary outside the breeding season are frequently observed flying in groups between feeding areas or to a roost for the night. Furthermore, most of the white species that tend to be solitary or semisocial outside the breeding season, become highly social at the start of each breeding season.

Being conspicuously white might be considered a disadvantage, especially for a small heron species, but we find that the smaller white species tend to be highly social (for example, *thula* and *garzetta*). By their size the larger all-white species would be relatively immune from attack. A rule of thumb might be formulated as follows: if you are a large heron species, it does not matter if you "go white," but if you are a small heron species that "goes white," you had better be highly social at the same time. The whiteness of the small, highly social species such as *thula* might be an adaptation for inter-individual contact. But why should a large, rather solitary species "go white?" Again, I believe the answer is conspicuousness. As previously stated, *albus* tends to feed alone, but moves in small groups, and during the breeding season it tends to be a highly social bird. In contrast, a flock of white *herodias* is unusual. These feed by themselves, even though dozens of birds may congregate on favorable feeding grounds. Roosting flights are usually performed alone. When the males are dancing and performing the Stretch and Snap Displays, however, they are very conspicuous, and their white coloration may function as a social signal. Unfortunately, this does not explain why the mainland *herodias* are blue.

In brief, I believe that the primary function of whiteness, at least in the highly social species, is related to conspicuousness in a variety of social situations.

SEXUAL DIMORPHISM AND PAIR FORMATION

The bitterns of the genus *Ixobrychus* are unique among the Ardeidae in that the sexes are readily separated by plumage. Because no detailed studies of the behavior of any member of this widespread genus have been made, the significance of this plumage dimorphism remains obscure. These small bitterns are typically solitary or semisocial, relatively silent, and they tend to inhabit *Phragmites, Typha,* and similar vegetation. The sexual differences in coloration may act as immediate visual sex recognition markers in these dense habitats.

Other heron species show noticeable, but not striking, sexual dimorphism during pair formation. Recall that in every species studied in detail, the males had much more brilliantly colored soft-parts and more lustrous plumage than did the females (see, for example, Table 9).

Not only do the sexes differ in the brilliance of their soft-part coloration and general plumage luster, but they also differ strikingly in behavior. Table 9 lists a number of behavioral differences between the sexes of *Butorides virescens*. There are striking sexual differences as to the time of appearance, orientation, frequency, and intensity of the performance of displays, and I believe that the greater brilliance of the males is correlated with such differences in behavior.

Two trends are evident in the pair formation of the herons studied so far: (1) a stationary male displays while a mobile female roams from one male to another; and (2) one male, with a rather mobile territory (at least during the early stages of pair formation) displays while a number of conspecific individuals gather to watch. With the type one species (*cinerea*, *virescens*, for example), one male and one female are typically involved, but with the type two species (for example, *thula* and *nycticorax*), communal displays are characteristic. However, the composition of the "spectators" with respect to sex in type two is not as yet clear.

In either type it is the "fixed" male which is on display; hence, the greater brilliance of his coloration and increased frequency and intensity of his displays tend to attract a female to his nest or nest site.

COLORATION OF SOFT-PARTS

According to Meiklejohn (1952), the Emperor Frederick II (1194–1250), in his book "De Arte Venandi cum Avibus," was the first to note the striking changes in color of the soft-parts of *Ardea cinerea* during the breeding season. Since Frederick's time a host of observers have seen and recorded these unusual changes in many heron species.

As I have said, I believe that the primary function of such color changes is to emphasize certain movements of the displaying bird. The color changes occur only at the onset of each breeding season, and colors fade rapidly during the period of egg-laying. A recrudescence of such brilliant coloration occurs prior to the laying of a second clutch (Lorenz, in Allen and Mangels, 1940; Noble and Wurm, 1940), or before renewal of pairing after the destruction of the nest (Binsbergen, 1941). Brilliant soft-part colors may function chiefly to emphasize sexual displays, their role in hostile situations possibly being secondary. Hostile displays are common outside the breeding season, especially among the social species (Snowy Egret and Black-crowned Night Heron, for example), but the soft-parts at this time are dull. Furthermore, many individuals begin to set up territories when their soft-parts are still dull or changing, but when they begin to display sexually, the soft-parts are always brilliant.

Because the majority of displays, especially the Stretch Display, are shown more frequently and more intensively by the male, it is reasonable to assume that the earlier acquisition and greater brilliance of the male's coloration is correlated with his greater activity, especially during sexual displays. Noble and Wurm (1942) and Lowe feel that the function of such brilliant colors is to enhance the displays, but these authors do not assign a primary role to brilliant soft-part colors in sexual displays and a secondary role in hostile displays as I do. However, my suggestions are speculative, and final conclusions must await experimentation.

As to the physiological basis for such changes in the coloration of the soft-parts, Noble and Wurm demonstrated that an androgen is responsible. When both sexes of the Back-crowned Night Heron were treated with testosterone propionate, the experimental birds showed a sharp increase in black pigment deposition in the lores and buccal cavity, darkening of the bill, and extreme vascularization of the legs. Full breeding coloration was induced in gonadectomized birds after treatment with testosterone propionate.

Noble and Wurm expressed the belief that the darkening of the lores was due to an "enormous increase in pigment below the epidermis," and that autonomic effects on the capillaries of the loral region played little or no role in such color changes. However, the melanins responsible for the black coloration of the lores of *nycticorax* are not the only pigments involved in such color changes in other heron species. For example, the yellow lores of *thula* and the red bill of *cinerea* and *herodias* are probably due to a combination of carotenoid pigments and some degree of vascularization. The effects of androgens on the deposition of carotenoids in these herons is not known.

With respect to the rich, rosy-red color of the legs and feet of the breeding Black-crowns, Noble and Wurm concluded that extensive vascularization, not pigment deposition, was responsible. When the integuments of the legs of experimental and control birds were compared, these authors observed that there was an increase in the size and number of the superficial blood vessels, together with an increased activity of the epidermal cells of the treated birds, but the number of pigment cells was the same in both controls and experimentals. Experimental proof is lacking, but I suspect that the slow changes in the coloration of the feet of such a species as *thula* may be a reflection of some pigment deposition and autonomic effects.

My observations on the sudden changes in coloration of the lores of Snowy Egrets during hostile clashes seem to indicate a combination of pigment activity and direct effects (autonomic) on the blood vessels of the loral region. During an intense hostile clash, the lores of the threatening egret changed rapidly from yellow to orange, and then finally to a brilliant scarlet; however, once the clash was ended, the lores faded rapidly to their original yellow coloration. Such observations clearly indicate to me that there is a complex interaction of pigment deposition, capillary development, and immediate autonomic effects. For a more detailed discussion of such autonomic effects, see Morris (1956). Noble and Wurm (1940:847) state, "It is probable that the androgen secreted by the night heron during the breeding season has a multiplicity of effects on the body." These same authors noted that rough handling of the Black-crowns frequently resulted in temporary color changes in the legs and feet of those birds that had been treated with testosterone propionate. The "blush" of the Little Bittern's (*Ixobrychus minutus*) legs is another example of such rapid color changes brought about by excitation (George Edward, in Lowe, 1954).

Because the head of a threatening heron is typically directed frontally toward the opponent, such *rapid* changes in lore and bill color may have a signal function in hostile encounters. Iris color changes that take place during intense hostile clashes (Green Heron, for example) may also act as "short range" social signals.

An additional function of soft-part color changes may be their role as supplementary isolating mechanisms. Among the heron species described in this paper no two that breed sympatrically have the same combination of breeding season colors. Such color differences, acting as visual signals, may function as one of many reinforcements in a chain-like series of reproductive isolating mechanisms.

Although too few species have been studied to formulate generalizations about the evolution of soft-part color changes, there appears to be a trend in all-white species toward *bright* colors in the bill and lores (for example, *albus* and *thula*), while in dark-bodied forms, the trend is to develop more subdued bill and loral coloration (*virescens* and *nycticorax*). For example, the all-white Cattle Egret has a brilliant red bill and lores (Tucker, 1936; Peterson *et al.*, 1954, among others). Adults of North American *albus*, which are all-white, have a brilliant orange bill and bright, lime-green lores, while those of the white *thula* have scarlet lores and jet black bills. In contrast, such dark-bodied species as *virescens* have black bills and black or bluish-black lores. A trend toward *contrast* in coloration is evident in a species such as the Black-crown. This species is dark above (crown and mantle), and it has dark soft-part colors during the breeding season on the bill and lores; however, the underparts of breeding *nycticorax* are pure white, and the legs and feet turn a brilliant rosy-red or salmon-pink. This trend toward contrasting colors is also shown by breeding Green Herons, which have dark bodies but brilliant coral-orange legs and feet.

Failure to take into account the changes in soft-part coloration that occur at the onset of the breeding season may lead to taxonomic difficulties, especially with regard to studies on intra-specific variation (Van Tyne, 1950:5; and Bernis, 1956).

SIZE AND ACTIVITY

Although generalizations based on the comparison of a limited number of species are always dangerous, one trend is the correlation between size and activity.

The smaller North American species, such as *virescens, thula,* and *rufescens,* all have a number of aerial displays, both hostile and sexual. For example, Flying Around, Pursuit Flights, Circle Flights, and Flap Flight Displays are very common in the courtship of *virescens,* while Aerial Stretch Displays and Jumping Over Displays are common in *rufescens* and *thula.* In contrast, the bulk of the displays, both hostile and sexual, of the larger species (*herodias* and *albus*) are mostly non-aerial in nature. In fact, many of the displaying males of the larger species perform mainly from one site—the nest platform or the crown of the nest tree.

The active hostile behavior of the smaller species (Full Forward Displays of *rufescens* and *thula*) is in sharp contrast with the relatively inactive threat behavior of the larger species. Two threatening individuals of *herodias* wade toward one another slowly, but *thula* and *rufescens* rush headlong to the attack. Threatening Black-crowns, which are stocky birds, typically stalk their opponents methodically, while Green Herons frequently attack a territory trespasser by flying directly toward the intruder.

The Stretch Display of *herodias* is performed simply and slowly, but the small *thula* pumps its head up and down vigorously, *rufescens* shows violent head-tossing during its Stretch Displays, and *virescens* actively sways from side to side. None of the larger herons have aerial Stretch Displays, but the smaller species show highly ritualized aerial variants of this sexual display. If a displaying male of one of the larger species is approached by the female, the male usually threatens while remaining stationary, but the typical response of one of the smaller species is active pursuit of the female.

In general, the feeding behavior of the larger species is rather inactive, Stand and Wait and Wade or Walk Slowly being the principal feeding methods. The techniques used in feeding by the smaller species (*thula* and *rufescens*) are in sharp contrast, such Disturb and Chase methods as Open Wing, foot-stirring, and Canopy Feeding being most prominent. However, a noticeable exception to the rule of small-active, large-inactive, is to be found in the Green Heron. This species is typically a slow feeder, Stand and Wait being its principal method of securing food. I believe that the inactive feeding behavior of *virescens* is one of several characters that link the Green Heron to the bitterns (*Ixobrychus*). The relatively inactive feeding methods of *Ixobrychus* may be prescribed by the dense nature of its typical habitat. The Black-crowned Night Heron is an inactive feeder, but I place this stocky species with the larger herons with respect to size.

After reading the many recent contributions on the comparative ethology of the gulls and their allies (family Laridae), it is evident to me that here also the small-bodied species are much more active than their large-bodied relatives. For example, the gulls of the *Hydrocoloeus* group (e.g., *Larus ridibundus* Linnaeus and *L. minutus* Pallas) show more active and more frequent aerial activities than do the larger species of the typical *Larus* group (e.g., *L. argentatus* and *L. marinus* Linnaeus). For details, see especially Moynihan (1955c, 1956, 1958) and Tinbergen (1953). Rensch (1954) should be consulted for a discussion of the relation between body size and central nervous functions.

COMPARATIVE BEHAVIOR CHART

The following chart is a summary of existing knowledge of the displays and related activities of ten North American herons. An "X" denotes the occurrence of a particular display or activity, while a question mark (?) signifies that information is unavailable or that descriptions in the literature are not detailed enough for comparative purposes. The numbers following certain question marks refer to available published descriptions. The herons are referred to by the following abbreviations:

GH—Green Heron, *Butorides virescens*
GB-GW—Great Blue and Great White Herons, *Ardea herodias*
CoE—Common (American) Egret, *Casmerodius albus*
RE—Reddish Egret, *Dichromanassa rufescens*
SE—Snowy Egret, *Leucophoyx thula*
LB—Little Blue Heron, *Florida caerulea*
LH—Louisiana Heron, *Hydranassa tricolor*
CaE—Cattle Egret, *Bubulcus ibis*
BC—Black-crowned Night Heron, *Nycticorax nycticorax*
YC—Yellow-crowned Night Heron, *Nyctanassa violacea*

Behavior	GH	GB-GW	CoE	RE	SE	LB	LH	CaE	BC	YC
Spiral descent flight	x	x	x	x	x	x	x	?	?	?
Hover over intruder at nest						?	?	?	x	?
Do not hover intruder	x	x	x	x	x	?	?	?		?
Fly directly down to roost	x							?		
Walk-jump down to roost		x	x	x	x	x	x	?	x	x
Underwing preening	x	x	x	x	x	x	x	?	x	?
Direct head-scratching	x	x	x	x	x	x	x	x	x	?
Shaking-out plumage	x	x	x	x	x	x	x	?	x	x
Sun-bathing	x	x	x	x	x	?	x	?	?	?
Bittern Stance	x	?	?	?	x	?	?1	?	?	?
Alert postures	x	x	x	x	x	x	x	?	x	?
Alarm call	x	x	x	x	x	x	x	x	x	x
Feeding techniques:										
Stand and Wait	x	x	x	x	x	x	x	x	x	x
Wade or Walk Slowly	x	x	x	x	x	x	x	x	x	x
Wing-flicking	?	x	x	x	x	x	x	?	?	?
Open Wing	?	?	?	x	x	x	x	?	?	?
Underwing	?	?	?	x	x	?	x	?	?	?
Canopy	?	?	?	x	?	?	?	?	?	?
Head-tilting	?	?	?	x	?	?	?	?	?	?
Foot-stirring	?	?	?	x	x	?	x	?	?	?
Foot-paddling	?	?	?	?	x	?	?	?	?	?
Hovering-stirring	?	?	?	x	x	?	?	?	?	?
Diving	x	x	?	?	?	?	?	?	x	?
Hovering	?	x	?	x	x	x	?	?	x	?
Onset of breeding protracted		x	x	x	x	x	?	?	x	x
Onset of breeding on arrival	x									

(Continued on p. 152)

COMPARATIVE BEHAVIOR CHART (Continued).

Behavior	GH	GB-GW	CoE	RE	SE	LB	LH	CaE	BC	YC
Non-breeding soft-part colors	x	x	x	x	x	x	x	x	x	x
Breeding soft-part colors	x	x	x	x	x	x	x	x	x	?
Establishment of Territory	x	x	x	x	x	?	?	?	x	?
Advertising call	x	?2	x	x	x	?	?	?	x	?
Flying Around	x	?	x	x	x	x	x	?	?	?
Pursuit Flight	x	?	x	x	x	x	x	?	?	?
Supplanting Attack	x	?	x	x	x	?	?	?	x	?
Nipping in flight	x	?	x	x	x	?	?	?	?	?
Crest-raising	x	x	x	x	x	x	x	?	x	x
Tail-flipping	x	x	x	x	x	x	x	?	x	x
Withdrawn Crouch	x	?	?	?	?	?	?	?	?	?
Forward Display	x	x	x	x	x	?	?	?	x	x
Full Forward Display	x	x	x	–	–	?	?	?	x	?
Upright Display	–	x	x	x	x	x	x	?	x	?
Aggressive Upright	–	x	x	x	x	?	?	?	–	?
Attack call	x	x	x	x	x	x	x	?	x	x
Stiff-necked Upright	x	?	?	?	?	?	?	?	?	?
Fighting	x	x	x	x	x	x	x	?	x	x
Bill-snapping	x	x	?	x	x	x	x	x	x	x
Feather-nibbling	x	x	x	x	x	x	x	x	x	x
Peering Down Display	?	?	?	x	?	?	?	?	?	?
Typical Snap Display	x	x	?	?	x	?3	?	?	x	?
Bowing or bobbing Snap	x	x	?	x	x	?3	?	?	?	?
Extended Stretch	x	?	?	?	?	?3	?	?	?	?
Stationary Stretch	x	x	?	x	x	?3	?4	?	x	?5
Aerial Stretch	?	?	?	x	x	?	?	?	?	?
Circle Stretch	?	?	?	x	?	?	?	?	?	?
Dancing Ground Display	?	x	x6	x	?	?	?	?	?	?
Gathering Ground Display	?	?	?	?	?	?	?	?	x	?
Circle Flight	x	x	x	x	x	?	?	?	?	?
Crooked Neck Flight	x	?	?	?	?	?	?	?	?	?
Flap Flight	x	?	?	?	?	?	?	?	?	?
Twig-passing	x	x	x	x	x	x	x	x	x	x
Jumping Over	?	?	?	x	x	?	?	?	?	?
Tumbling Flight	?	?	?	?	x	?	?	?	?	?

1—Palmer, 1909; 2—Cottrille and Cottrille, 1958; 3—Meanley, 1955; 4—Huxley (in Bent, 1926); 5—Nice, 1929; 6—Audubon (in Bent, 1926).

LITERATURE CITED

ALLEN, A. A. 1929. Green Heron's story. Bird-Lore, **31**:289–298.

——— 1934. American Bird Biographies. Ithaca, New York: Comstock, 238pp.

——— 1951. Stalking Birds with Color Camera. Washington, D.C.: Nat'l. Geog. Soc., 328pp.

ALLEN, ROBERT PORTER. 1942. The Roseate Spoonbill. New York: Nat'l. Aud. Soc., Res. Rep. No. 2, 142pp.

——— 1954–55. The Reddish Egret: bird of colors and contrasts. Aud. Mag., **56**:252–255; **57**:24–27.

ALLEN, ROBERT P., AND FREDERICK P. MANGELS. 1940. Studies of the nesting behavior of the Black-crowned Night Heron. Proc., Linn. Soc. N.Y., Nos. 50, 51:1–28.

AMERICAN ORNITHOLOGISTS' UNION. 1957. Check-List of North American Birds. Fifth Ed. Baltimore: Amer. Ornith. Union, 691pp.

ANDREW, R. J. 1956–57. The aggressive and courtship behaviour of certain emberizines. Behaviour, **10**:255–308.

ARMSTRONG, E. A. 1946. The coloration of sea birds. Why does white plumage predominate? Birds of Britain, **2**:15–19.

AUDUBON, JOHN JAMES. 1840. The Birds of America. New York: J. J. Audubon.

AUDUBON, JOHN JAMES (see Bent, 1926).

AYRES, THOMAS. 1878. Additional notes on the ornithology of Transvaal. Ibis (4th Ser.), II(7):281–301.

BAERENDS, G. P., AND J. M. BAERENDS-VAN ROON. 1950. An Introduction to the Study of the Ethology of Cichlid Fishes. Behaviour, Suppl. I, 242pp.

BAIRD, S. F., T. M. BREWER, AND R. RIDGWAY. 1884. The Water Birds of North America. Vol. I. Boston: Little, Brown & Co., 537pp.

BARKER, SAMUEL H. 1901. Does the Green Heron fish in deep water? Bird-Lore, **3**:141.

BASTOCK, MARGARET, AND AUBREY MANNING. 1955. The courtship of *Drosophila melanogaster*. Behaviour, **8**:85–111.

BEETHAM, BENTLEY. 1910. The Home Life of the Spoonbill, the Stork, and Some Herons. London: Witherby, 47pp.

BENT, ARTHUR CLEVELAND. 1926. Life Histories of North American Marsh Birds. Bull., U.S.N.M., **135**:1–490.

BERNIS, F. 1956. Sobre la Garceta (*Egretta*) de color negruzco aparecida en Doñana. Ardeola, **3**:93–114.

BINSBERGEN, NOL. 1941. Uit Neerlands Vogelleven. (see Brit. Birds, **42**:48–50).

BLAKE, CHARLES H. 1948. More data on the wing flapping rates of birds. Condor, **50**:148–151.

BOCK, WALTER J. 1956. A generic review of the Family Ardeidae (Aves). A.M.N.H., Nov. No. 1779:1–49.

BROOKS, W. SPRAGUE. 1923. An interesting adaptation. Auk, **40**:121–122.

BROWN, A. GRAHAM. 1949. Notes on some birds of the Whitsunday Group, Queensland. Emu, **49**:44–49.

BURLEIGH, THOMAS D. 1929. Notes on the bird life of northwestern Washington. Auk, **46**:502–519.

CARPENTER, CHARLES KNAPP. 1948. An early Illinois record of "Cory's Least Bittern." Auk, **65**:80–85.

CHRISTY, BAYARD H. 1929. Notes. Cardinal, **2**:129.

CONDER, P. J. 1949. "Individual distance." Ibis, **91**:649–655.

COTT, HUGH B. 1947. The edibility of birds. Proc. Zool. Soc. Lond., **116**: 371–524.

COTTRILLE, W. POWELL, AND BETTY DARLING COTTRILLE. 1958. Great Blue Heron: Behavior at the Nest. Misc. Pub., Mus. Zool., U. Michigan, No. 102, 15pp.

COWLES, RAYMOND B. 1930. Notes on the nesting of the African Green Heron (*Butorides atricapilla*) in Natal. Auk, **47**:465–470.

CRAIK, K. J. W. 1944. White plumage of sea-birds. Nature, **153**:288.

CRUICKSHANK, ALLAN D. 1942. Birds Around New York City. A.M.N.H., Handbook Ser., No. 13, 489pp.

CRUICKSHANK, HELEN. 1948. On the trail of the heron with the golden slippers. Aud. Mag., **50**:138–148.

DAANJE, A. 1950. On locomotory movements in birds and the intention movements derived from them. Behaviour, **3**:48–98.

DARWIN, CHARLES. 1890. The Descent of Man. 2nd Ed. London: Murray.

DELACOUR, J. 1946. Under-wing fishing of the Black Heron, *Melanophoyx ardesiaca*. Auk, **63**:441–442.

DELACOUR, J., AND ERNST MAYR. 1945. The Family Anatidae. Wilson Bull., **57**:3–55.

——— 1946. Supplementary notes on the Family Anatidae. Wilson Bull., **58**:104–110.

DILGER, WILLIAM C. 1956. Hostile behavior and reproductive isolating mechanisms in the avian genera *Catharus* and *Hylocichla*. Auk, **73**:313–353.

FINN, FRANK. 1919. Bird Behaviour. New York: Dodd, Mead & Co., 363pp.

FISHER, G. CLYDE. 1933. Nature's Secrets. Part I. New York: Univ. Society, Inc., 462pp.

FORBUSH, EDWARD HOWE. 1929. Birds of Massachusetts. Part I. Mass. Dep't. Agricul., 481pp.

FRISCH, OTTO VON. 1957. Mit einem Purpurreiher verheiratet. Z. Tierpsychol., **14**:233–237.

GLENISTER, A. G. 1951. The Birds of the Malay Peninsula, Singapore and Penang. London: Oxford Univ. Press, 282pp.

GRISCOM, LUDLOW. 1923. Birds of the New York City Region. A.M.N.H. Handbook Ser., No. 9, 400pp.

——— 1926. The ornithological results of the Mason-Spinden Expedition to Yucatan. A.M.N.H. Nov., No. 235:1–19.

HAUSER, DORIS C. 1957. Some observations on sun-bathing in birds. Wilson Bull., **69**:78–90.

HAVERSCHMIDT, FR. 1948a. A feeding habit of the Snowy Egret. Wilson Bull., **60**:187.

——— 1948b. Bird weights from Surinam. Wilson Bull., **60**:230–239.

——— 1949. The Life of the White Stork. Leiden: E. J. Brill, 96pp.

——— 1953. Ardeidae with red legs. Ibis, **95**:699.

HAWBECKER, ALBERT C. 1949. Green Heron feeds on goldfish. Auk, **66**:78–79.

HEDIGER, H. 1942. Wildtiere in Gefangenschaft. Transl. as Wild Animals in Captivity, London, 1950.

HEINROTH, OSKAR. 1911. Beiträge zur Biologie, namentlich Ethologie und Psychologie der Anatiden. Verhandl. V. Int. Ornith.-Kongr., Berlin, 1910: 589–702.

——— 1930. Über bestimmte Bewegungsweisen bei Wirbelttieren. Stizungsber. Ges. Naturf. Freunde: 333–342.

HEINROTH, OSKAR, AND MAGDALENA HEINROTH. 1929. Die Vögel Mitteleuropas. Berlin-Lichterfelde: Hugo Bermuhler Verlag.

HICKEY, JOSEPH J. 1937. Notes on leg colors of white herons. Condor, **39**:131.

HINDE, R. A. 1953a. The conflict between drives in the courtship and copulation of the Chaffinch. Behaviour, **5**:1–31.

——— 1953b. Appetitive behaviour, consummatory act, and the hierarchical

organisation of behaviour—with special reference to the Great Tit (*Parus major*). Behaviour, **5**:189–224.
——— 1956a. Ethological models and the concept of "drive." Brit. Jour. Philos. Sci., **6**:321–331 (No. 24).
——— 1956b. The biological significance of the territories of birds. Ibis, **98**: 340–369.
HINDWOOD, K. A. 1933. The Green-backed Mangrove Heron. Parts I & II. Emu, **33**:27–43; **33**:97–102.
HOBBS, JOHN N. 1957. Feeding habits of some water birds. Emu, **57**:216.
HOLSTEIN, VAGN. 1927. Fiskeheiren. Copenhagen: G. E. C. Gads Vorlag, 98pp.
HOWELL, ARTHUR H. 1932. Florida Bird Life. New York: Coward-McCann, 579pp.
HUXLEY, J. S. 1922. Preferential mating in birds with similar coloration in both sexes. Brit. Birds, **16**:99–101.
JACKSON, F. J. 1938. The Birds of Kenya Colony and the Uganda Protectorate. London & Edinburgh: Gurney & Jackson, 3 vols., 1592pp.
JONES, LOMBARD CARTER. 1936. Snowy Egret in Massachusetts. Auk, **53**: 438–439.
KALTER, LOUIS B. 1932. A comparatively tame Eastern Green Heron. Auk, **49**:342.
KOENIG, OTTO. 1952. Ökologie und Verhalten der Vögel des Neusiedlersee-Schilfgürtels. J. f. Ornith., **93**:207–289.
LACK, DAVID. 1953. The Life of the Robin. London: Penguin Books, 240pp.
——— 1954a. The stability of the heron population. Brit. Birds, **47**:111–121.
——— 1954b. The Natural Regulation of Animal Numbers. London: Oxford Univ. Press, 343pp.
LORENZ, KONRAD. 1938. A contribution to the comparative sociology of colonial-nesting birds. Proc. VIII Int. Ornith. Congr., Oxford, 1934:207–218.
——— 1941. Vergleichende Bewegungsstudien an Anatinen. J. f. Ornith., **89**, Sonderheft, 194–293.
——— 1955. Morphology and behavior patterns in closely allied species. In: "Group Processes," Trans. 1st Conf., 1954:168–220.
LOVERIDGE, ARTHUR. 1922. Notes on East African birds (chiefly nesting habits and stomach contents). Proc. Zool. Soc. Lond., 837–862.
LOWE, FRANK A. 1954. The Heron. London: Collins New Naturalist Monograph No. 11, 177pp.
MARLER, P. 1956. Behaviour of the Chaffinch, *Fringilla coelebs*. Behaviour, Suppl. V, 184pp.
MAYR, ERNST. 1956. Is the Great White Heron a good species? Auk, **73**:71–77.
MAYR, ERNST, AND DEAN AMADON. 1941. Geographical variation in *Demigretta sacra* (Gmelin). A.M.N.H. Nov., No. 1144:1–11.
MCILHENNY, E. A. 1936. Unusual feeding habits of some of the Ardeidae. Auk, **53**:439–440.
MCLEAN, D. D. 1930. The speed of flight in certain birds. Gull, **12**:1–2.
MEANLEY, BROOKE. 1955. A nesting study of the Little Blue Heron in Eastern Arkansas. Wilson Bull., **67**:84–99.
MEIKLEJOHN, M. F. M. 1952. Ardeidae with red soft parts. Ibis, **94**:545.
MEINERTZHAGEN, R. 1949. Notes on Saudi Arabian birds. Ibis, **91**:465–482.
MEYERRIECKS, ANDREW J. 1957a. Louisiana Heron breeds in New York City. Wilson Bull., **69**:184–185.
——— 1957b. Field observations pertaining to the systematic status of the Great White Heron in the Florida Keys. Auk, **74**:469–478.
——— 1959. Foot-stirring feeding behavior in herons. Wilson Bull., **71**: 153–158.

MEYERRIECKS, ANDREW J., AND ROBERT MEYERRIECKS. 1958. Great White Heron. Nat. Hist., **67**:52–56.

MITCHELL, CATHERINE A. (see Bent, 1926)

MONSON, GALE. 1951. Great Blue Heron killed by bobcat. Wilson Bull., **63**:334.

MORRIS, DESMOND. 1956. The feather postures of birds and the problem of the origin of social signals. Behaviour, **9**:75–113.

MOYNIHAN, M. 1955a. Remarks on the original sources of displays. Auk, **72**: 240–246.

——— 1955b. Types of hostile display. Auk, **72**:247–259.

——— 1955c. Some Aspects of Reproductive Behaviour in the Black-headed Gull (*Larus ridibundus ridibundus* L.) and Related Species. Behaviour, Suppl. IV, 201pp.

——— 1956. Notes on the behavior of some North American gulls. I. Aerial hostile behavior. Behaviour, **10**:126–178.

——— 1958. Notes on the behavior of some North American gulls. II. Non-aerial hostile behavior of adults. Behaviour, **12**:95–182.

NICE, MARGARET MORSE. 1929. Some observations on the nesting of a pair of Yellow-crowned Night Herons. Auk, **46**:170–176.

NOBLE, G. K. 1939. The role of dominance in the social life of birds. Auk, **56**: 263–273.

NOBLE, G. K., AND M. WURM. 1940. The effect of testosterone propionate on the Black-crowned Night Heron. Endocrinology, **26**:837–850.

——— 1942. Further analysis of the social behavior of the Black-crowned Night Heron. Auk, **59**:205–224.

——— AND A. SCHMIDT. 1938. The social behavior of the Black-crowned Night Heron. Auk, **55**:7–40.

PALMER, WILLIAM. 1909. Instinctive stillness in birds. Auk, **26**:23–36.

PEMBERTON, J. R. 1922. The Reddish Egrets of Cameron County, Texas. Condor, **24**:3–12.

PETERSON, ROGER TORY, GUY MOUNTFORT, AND P. A. D. HOLLOM. 1954. A Field Guide to the Birds of Britain and Europe. London: Collins, 318pp.

PHILLIPS, W. W. A., AND R. W. SIMS. 1958. Two new races of birds from the Maldive Archipelago. Bull. Brit. Ornith. Club, **78**:51–53.

PORTIELJE, A. F. J. 1926. Zur Ethologie bezw. Psychologie von *Botaurus stellaris*. Ardea, **15**:1–15.

POUGH, RICHARD H. 1951. Audubon Water Bird Guide. New York: Doubleday & Co., 352pp.

RAND, A. L. 1936. The Distribution and Habits of Madagascar Birds. Summary of the Field Notes of the Mission Zoologique Franco-Americaine a Madagascar. Bull., A.M.N.H., **72**:143–499.

——— 1956. Foot-stirring as a feeding habit of Wood Ibis and other birds. Amer. Mid. Nat., **55**:96–100.

RENSCH, BERNHARD. 1954. The relation between the evolution of central nervous functions and the body size of animals. In: "Evolution as a Process," London: Allen & Unwin, Ltd., 367pp.

RICE, DALE W. 1954. Symbiotic feeding of Snowy Egrets with cattle. Auk, **71**:472–473.

——— 1956. Dynamics of range expansion of Cattle Egrets in Florida. Auk, **73**:259–266.

ROBERTS, THOMAS S. 1936. The Birds of Minnesota. Vol. I. Minneapolis: Univ. Minnesota Press, 718pp.

SAUNDERS, ARETAS A. 1926. The summer birds of central New York marshes. Roosevelt Wildlife Bull., **3**:335–475.

SCHÜZ, E. 1942. Bewegungsnormen des Weissen Storchs. Z. Tierpsychol., **5**: 1–37.

SCOTT, W. E. D. 1890. On birds observed at the Dry Tortugas during parts of March and April, 1890. Auk, **7**:301–314.

SEIBERT, HENRI C. 1951. Light intensity and the roosting flight of herons in New Jersey. Auk, **68**:63–74.

SHERWOOD, MARY P. 1957. Camping by a bird stage in the Everglades. Florida Naturalist, **30**:11–18.

SIMMONS, GEORGE FINLAY. 1925. Birds of the Austin Region. Austin: Univ. Texas Press, 387pp.

SIMMONS, K. E. L. 1957. The taxonomic significance of the head-scratching methods of birds. Ibis, **99**:178–181.

SLIPP, JOHN W. 1942. Green Heron at Tacoma, Washington. Condor, **44**:35–36.

SMITH, J. B. 1894. "High Jinks" of the Great Blue Heron. Nidiologist, **2**:10–11.

SMYTHIES, BERTRAM E. 1953. The Birds of Burma. Edinburgh: Oliver & Boyd, 668pp.

SPRUNT, ALEXANDER, JR., AND E. BURNHAM CHAMBERLAIN. 1949. South Carolina Bird Life. Columbia, S. C.: Univ. South Carolina Press, 585pp.

STANFORD, J. K. 1936. Coloration of soft parts in the Herodiones during the breeding season. Brit. Birds, **30**:136.

STONE, WITMER. 1937. Bird Studies at Old Cape May. Philadelphia: Delaware Valley Ornith. Club, Vol. I, 520pp.

STORER, JOHN H. 1948. The Flight of Birds. Cranbrook Inst. Sci., Bull. 28, 94pp.

SUTTON, GEORGE MIKSCH. 1936. Food capturing tactics of the Least Bittern. Auk, **53**:74–75.

THORPE, W. H. 1956. Learning and Instinct in Animals. Cambridge: Harvard Univ. Press, 493pp.

TINBERGEN, N. 1952. "Derived" activities; their causation, biological significance, origin, and emancipation during evolution. Quart. Rev. Biol., **27**: 1–32.

——— 1953. The Herring Gull's World. London: Collins New Naturalist Monograph, No. 9, 255pp.

——— 1954. The origin and evolution of courtship and threat display. In: "Evolution as a Process," London: Allen & Unwin, 367pp.

TOWNSEND, CHARLES W. 1928. The song of the Green Heron. Auk, **45**:498–499.

TUCKER, B. W. 1936. The colouring of the soft parts of the Buff-backed Heron. Brit. Birds, **30**:70–73.

VAN TYNE, JOSSELYN. 1950. Bird notes from Barro Colorado Island, Canal Zone. Occ. Papers, Mus. Zool., Univ. Michigan, No. 525:1–12.

VERWEY, JAN. 1930. Die Paarungsbiologie des Fischreihers. Zool. Jahrb., Abt. Allgem. Zool. Physiol. Tiere, **48**:1–120.

WAARD, S. DE. 1937. De snapbeweging van de blauwe reiger. Lev. Nat., **41**: 159–160.

WALSH, LESTER L. 1929. Snowy Egret (*Egretta candidissima candidissima*) in northern New Jersey. Auk, **46**:536–537.

WARBURTON, FRED. 1948. Green Heron captures flying dragonflies. Auk, **65**:132.

WELLS, H. G., JULIAN S. HUXLEY, AND G. P. WELLS. 1929. The Science of Life. New York: Literary Guild, 1514pp.

WESTON, FRANCIS MARION (see Howell, 1932).

WETMORE, ALEXANDER. 1916. The speed of flight of certain birds. Condor, **18**:112–113.

WHEELOCK, IRENE G. 1906. Nesting habits of the Green Heron. Auk, **23**: 432–436.

WHITE, C. A. 1947. Night-heron alighting on water and swimming. Brit. Birds, **40**:314.

WILSON, ALEXANDER, AND CHARLES LUCIAN BONAPARTE. 1878. American Ornithology. One Volume Edition. Philadelphia: Porter & Coates.

WINTERBOTTOM, J. M. 1957. Some African Herons. African Wild Life, **11**: 100–105.

WITHERBY, H. F., F. C. R. JOURDAIN, N. F. TICEHURST, AND B. W. TUCKER. 1949. The Handbook of British Birds. Vol. III. London: H. F. & G. Witherby Ltd., 387pp.

WOOD, HAROLD B. 1933. Flight speed of some birds. Auk, **50**:452–453.

PLATE 1. Green Heron preening its upper neck feathers.

PLATE 2. Green Heron preening the underside of its wing.

PLATE 3. Green Heron scratching the side of its head directly.

PLATE 4. Green Heron shaking out its plumage.

Plate 5. Green Heron sleeping while brooding young in the nest.

Plate 6. The normal perch position of the Green Heron.

PLATE 7. Crest-raising by the Green Heron.

PLATE 8. Young Great White Heron preening its neck feathers.

PLATE 9. Young Great White Heron preening one of its primaries.

PLATE 10. Young Great White Heron sun-bathing.

Plate 11. Young Great White Heron giving the Forward Display in its nest.

PLATE 12. Reddish Egret showing Head-tilting while feeding.

PLATE 13. Canopy Feeding by the Reddish Egret.

PLATE 14. Canopy Feeding by the Reddish Egret.

PLATE 15. Snowy Egret preening the underside of its wing.